Sudoku Seventeen, by John Austin

Sudoku

Seventeen

A selection of 200 special, challenging puzzles with a step-by-step guide to their solution

John Austin, Ph.D.

Sudoku Seventeen, by John Austin

Published by Enigma Scientific Publications

First published in England in 2014 by Enigma Scientific Publications, Berkshire, UK
http://www.enigmascientific.co.uk

First print version 12 December, 2014.
ISBN number 9781849146036

CreateSpace version 26 October 2015.
ISBN number 9781518789908

Contents

Sudoku Seventeen, by John Austin

Sudoku Seventeen, by John Austin

1.Introduction

The solving of Sudoku puzzles has become a leisure activity that hardly needs any introduction. Most puzzles that are published in newspapers and magazines range in difficulty from easy to challenging and perhaps higher. They have varying characteristics in which the clues or given numbers, or poles as I refer to them, vary in position and in such a way as to create pleasing shapes or symmetries at the whim of the composer. Typically puzzles have 23-27 poles. You might think that 23 is near the lower end of the possible number, as I once did. Don't forget that we are looking for solutions to the puzzle which are unique. By reducing the number of poles we run the risk that the puzzle will admit many solutions. For example, trivially, if the puzzle contained just a few poles (Figure 1.1) the number of solutions would be immense and the puzzle would no longer strictly be a Sudoku puzzle in having a unique solution. I have come across puzzles with the number of poles in the lower 20s which have a handful of solutions and therefore strictly break the rules. These are a headache to solve as it is difficult to know when exactly to quit and use a trial and error mode of solution. There is a very special class of Sudoku puzzles which have just 17 poles, yet remarkably still have unique solutions. Although it was conjectured that 17 was the minimum number of poles that could provide a unique solution, this was not proven until relatively recently[1]. In anticipation of this, a large collection of 17 pole Sudokus was made[2] and I will refer to each of these as Sudoku-17 puzzles.

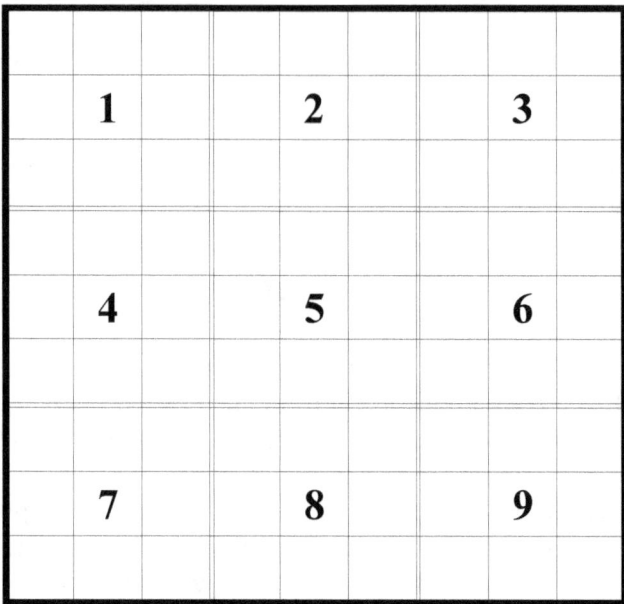

Fig. 1.1: The above rather silly "puzzle" has far too few poles to be considered an actual Sudoku puzzle.

In my previous book on Sudoku and related puzzles[3], I provided a step-by-step method for solving them. This book goes that little step further by providing a selection of over 200 Sudoku-17 puzzles with a guide to their solution. You might think that these Sudokus would be tremendously challenging and some are indeed very difficult. However, although there are no easy puzzles here, many 17-pole Sudokus are solvable with an extension of the techniques supplied in [3]. I personally found this quite eye-opening, as before trying them I thought that it would be a matter of hard grind and substantial amounts of trial and error would be needed to reach a solution. Moreover the extension of the usual techniques provides that extra little interest that perhaps experienced Sudoku solvers will look for. This book shows, I think, the rich tapestry of structure and form that occurs within a relatively trivial mathematical structure.

Incidentally, in solving Sudoku puzzles over the years

like me, many solvers will presumably start to pick up on the nuances of individual collections. When you obtain a collection of puzzles from one source, they often have particular characteristics that may be difficult to describe precisely. I believe that this reflects the brain's tendency to be able to see form and structure which is difficult to define. For example if you listen to a few bars of music by Mozart or Mahler or Shostakovic (which is more my style) or even music by popular artists, you can almost identify the artist straight away. The same is true of the visual arts, Constable, Turner, Picasso etc. Why is that? On a similar (but less esoteric!) note Sudoku-17 puzzles have their own individual and enjoyable style.

In reading this book I would suggest that you first read the remainder of this Chapter. It provides background information on the mathematical proof that "there is no 16-clue Sudoku". I then describe the puzzles selected for this work.

After reading Chapter 1, I would suggest that you attempt to tackle some of the puzzles selected. If you find these too difficult or if you want to compare my techniques with yours, you can go back to Chapter 2. There I solve several examples of different complexity. At the end of Chapter 2, I mention the world's most difficult Sudoku puzzle. The puzzle itself is presented at the beginning of the puzzle section and the solution with some hints are given at the beginning of the solution section. I would suggest that you leave this to last, but it is there to bring you down to earth, if you think the other puzzles are too easy! It does have certain characteristics in common with the other puzzles in this book, but it has 21 poles so it is not a Sudoku-17 puzzle.

1.1 There is no 16-Clue Sudoku

Although the above statement was conjectured, there was no proof until 2013[1]. The proof used a supercomputer to search through every possible 16-clue Sudoku and no solutions were found. There is a vast number of possible sudoku grids to

Sudoku Seventeen, by John Austin

check by brute force, but with a combination of methods: brute force (i.e. explicitly solving every puzzle) and reduction of Sudoku that are mathematically equivalent, the computer time could be reduced from 300,000 computer processor years to a mere 800 processor years! By comparison, a century simulation of a climate model typically takes about 500-1000 processor years. There was also strong, circumstantial evidence that there is no 16-clue Sudoku. The reasoning of McGuire et al.[1] was that one particular grid contained as many as 29 of the 17-pole Sudokus already known. Nonetheless, for the 16-pole Sudoku to exist there would need to be a sudoku puzzle which contains over 60 17-pole Sudokus to allow for the permutations of the 17-pole puzzle. Thus, although the computational method was a "null" result potentially open to error, all the evidence points to their conclusion being correct. Although the 16-pole puzzle doesn't exist, mathematical curiosities abound, with the 16-pole puzzle in Figure 1.2 (due to Royle) having just 2 solutions.

Ideally, one would have liked a mathematical proof rather than a process which partially relies on brute force. However, as noted in [1], mathematics alone, without computer aid, can demonstrate that a 7-clue puzzle does not have a unique solution. This is of course still too far removed from the known minimum number of clues to be useful information. Interestingly, a mathematical "proof" of the non-existence of a 16-clue Sudoku was published by a young mathematician in 2008[4], but this was subsequently shown to be incorrect, although I can't understand the German anyway.

McGuire et al.[1] go on also to summarise the smallest critical sets, as it is known, for other Sudoku grids of smaller size. For grid dimensions of 4, 6, 8 and 9 (standard Sudoku) the values found are now known to be 4, 8, 14 and of course 17. Mini-Sudoku, described in my previous Sudoku book[3] therefore requires at least 8 poles, although as we saw in [3], when the long diagonals are included as additional "rows" the number of poles can be reduced further.

Sudoku Seventeen, by John Austin

1								5
				3				
		2		4				
	3	4				7		
			2		6			1
2					5			
	7						3	
				1				

Figure 1.2: a 16-pole puzzle with just two solutions. This puzzle may be unique (apart from mathematical perturbations). Note that both 8 and 9 are missing from the poles.

1.2 Information per element

One issue of general interest to me and perhaps might be to you, is the information content in each element of the grid. For example, if you take a completed Sudoku grid and award it 81 points of information, *poi*, (1 point for each element) then in principle you could compare the information in each of the poles. You could argue that the grid only contains 72 *poi* as for example a trivial Sudoku puzzle is really already complete when it has all the numbers 1-8 in place as the 9th number follows by continuity. Instead I'll argue that the additional *poi* are spread amongst the completed elements. So, in a general puzzle with say 25 poles, each pole must have 81/25 = 3.24 *poi* on average. Some poles will have very little information while other poles will have more, in order to bring the average to 3.24. The process of solving takes

Sudoku Seventeen, by John Austin

advantage of the fact that the average pole has a *poi* value exceeding 1 and the idea is to propagate the additional information to other parts of the grid to determine the unknown elements. Interestingly, in some puzzles some poles are almost redundant. In other words, they could have been deduced from other information in the Sudoku puzzle. So these particular poles have low *poi* values. Whether you consider the value to be zero (because they can be entirely calculated by the other poles) or exactly 1.0 *poi* since you could argue that they can have no other influence on the grid. Perhaps this sort of ambiguity reflects the qualitative nature of these concepts. In any case, for a given puzzle each element does not have the same value. What this means is that if at some point a trial and error process is begun to solve the puzzle, it helps to be able to locate which elements have high *poi* values and to specify those values.

Now if we think about Sudoku-17 puzzles (Sudoku with 17 poles). Each pole could be thought of as having information $81/17 = 4.76$ *poi*. This is of course a lot higher (47%) than in the above example. Moreover we know that 17 is the minimum number of poles, so each pole is adding information to the solution other than its own grid value. Overall, the extra information that each pole contains is presumably related to the complexity of the structure that it forms with the other elements. Every pole counts, so in the solution, the solver can focus more on each pole to determine what that particular pole is contributing to the solution. From my perspective, the increased interaction between different elements of the grid provides one of the most stimulating aspects of Sudoku-17 puzzles. Of course the fact that Sudoku-17 puzzles can't be reduced further has a nice philosophical air to it. It means that the puzzles can be thought of as fundamental in some regards (like elements in the periodic table) and from a mathematical perspective the solution proceeds from a sort of set of first principles. Mathematicians like that!

1.3 Description of the Puzzles

Each of the puzzles presented here was kindly supplied as a text file from Gordon Royle of the University of Western Australia. Each of the puzzles was converted (laboriously!) to a Sudoku grid that I had prepared in OpenOffice. I have solved every one of the puzzles myself without resorting to machine intelligence and I have therefore been able to give a difficulty rating. Such ratings are in any case subjective, but my guidelines were approximately: under 30 min. -- 3* ; 30-45 min. -- 4*; over 45 min. -- 5*. I also found a small set of puzzles which typically took over an hour to solve (6*). I call these puzzles "atomic" for entertainment purposes! The puzzles rated as 4* are about the same level as the Times "fiendish", while the 5* are about the same level as the "Challenging" ones in the Metro newspaper. Apparently the Metro editors claim that some Sudoku solvers find the challenging examples noteworthy to the extent of writing in to boast of their solving prowess. I do not argue with that. I do not claim to be particularly clever, so the target times above must be considered as pretty soft!

Over the months that I have been doing these puzzles, my times have gradually improved. For this reason, I went back to all of the early puzzles that I found difficult (and had placed in categories 5 and 6) and reclassified them. I was able to solve them a lot faster than initially. Not because I had remembered the puzzles themselves, but rather because I have become familiar with several themes which started to repeat (see for example Chapter 2.3). As a result, my actual times towards the end of the puzzle solving period were about 20% less than the above times. Of course I had the occasional "bad day at the office" and I made a mess of some of the puzzles. I have attempted to correct for this. Overall, though, in puzzle rating I have tried to remain guided primarily by complexity. In the limited number of cases where I have needed to use iteration (trial and error) methods, the times include that used to explore all the potential options, thereby negating any influence that luck might have played in the time taken.

In the original list, the puzzles are mathematically inequivalent[2]. That is, they cannot be transformed into each other by permuting the 9 symbols 1-9 or rearranging the rows, columns or regions. With 49,151 puzzles available at the time of writing (October 2014), statistical arguments presented in [1] suggest that there are very few new Sudoku-17 puzzles yet to be found.

The above method of determining the difficulty level is obviously subjective. There have been many attempts by mathematicians to quantify the difficulty of a puzzle. The principle adopted is typically to award points as each element is resolved. For example, a singlet could score 1 point and a doublet 3 points and so on. A more difficult puzzle would have a higher preponderance of elements that are only solved at the quadruplet level. Those that remain, namely, those requiring trial and error might require special techniques to resolve and higher points are awarded accordingly. While this appears a fairly objective method, in practice it presupposes that each solver will approach the puzzle in the same manner and therefore arrive at the same score. The fact that puzzle difficulty is subjective indicates that this isn't the case, i.e., people approach a puzzle from different angles and therefore would sum up the difficulty points in a different way.

After the puzzle section, I also give a set of hints for the solution of some of the puzzles. I am assuming here that the reader has not taken the trouble of acquiring some of the advanced techniques that can be used to eliminate options for some elements, which are known in the Sudoku puzzling community as wings, swordfishes and so on. To be honest, I haven't learnt these advanced methods myself either, as they are used rarely, are difficult to apply and difficult to memorise. Of course the reader can always cheat by looking at the solutions, an option that wasn't open to me when I put this material together! You often see the availability of hints for newspaper Sudoku

Sudoku Seventeen, by John Austin

puzzles, but it always struck me as being a very expensive way of satisfying your curiosity. Hints are often charged at £1 at a time, which could be comparable to the cost of the newspaper! Anyway, I have never used hints by texting to a number, so I don't know what they look like. Instead I have written a brief remark for the puzzle in question: usually a place from which to start the trial and error procedure. For these places, I have tried to identified elements which seem to have high *poi* values. Also, accidents do happen, so I have included 10 pages of blank grids immediately after the hints. If you make a mistake from which there is no return, do what I do and start again from scratch with a fresh grid! Make sure, though, that you copy the puzzle down exactly. With a "normal" puzzle making a mistake will often lead to a puzzle being unsolvable. With these puzzles, the chances are that with a mistake you will end up with a slightly different puzzle which will have multiple solutions. This will, frankly, be quite confusing.

      The solutions of the puzzles follow in Chapter 4. They have all been tested for uniqueness. Of course you hardly need to know the solutions if you've solved the puzzle correctly yourself. These provide additional hints if needed, and although it was laborious to fill in all the Sudoku grids for publication, it just seems *traditional* somehow to give the answers!

References

[1] There is no 16-Clue Sudoku: Solving the Sudoku Minimum Number of Clues Problem, G. McGuire, B. Tugemann and G. Civario, 1 September 2013, http://arxiv.org/pdf/1201.0749v2.pdf, accessed 27 October 2014.
[2] Minimum Sudoku, Gordon Royle, http://staffhome.ecm.uwa.edu.au/~00013890/sudokumin.php, accessed 27 October 2014.
[3] World of Sudoku, John Austin, 2014, 216pp, ISBN 9781849145664
[4] Mindestens 17 mussen es sein (in German), Ariane Papke, Junge wissenschaft, 84, 24-31, 2008.

Sudoku Seventeen, by John Austin

Sudoku Seventeen, by John Austin

2. How to Solve 17-Pole Sudokus

Before tackling examples of the solution of puzzles with Moderate, Difficult and Very Difficult levels of difficulty, we need to present the terminology used here. This is reproduced from my earlier work[1] and for further details, see the Glossary, also reproduced from [1].

2.1 Some Jargon

It is useful at this point to introduce a few terms, which are supplemented with material in the Glossary. The Sudoku puzzle itself I often refer to as a *grid* and the individual squares as *elements*. The puzzle consists of typically 9 *rows* and 9 *columns* which are self-explanatory. The puzzle can also be thought of as containing 9 *regions* which are the groups of 3x3 elements which make up the puzzle. So, the goal of the puzzle is to complete each row, column and region so that each includes the numbers 1-9.

In the simplest situation I refer to a single incomplete element as a *singlet*. Similarly, doublets are elements which can only be one of two specific numbers. For example, you might be able to deduce that a given element must be either 4 or 5. Similarly it is useful to consider triplets e.g. 3,4,5 and quadruplets e.g. 2,3,4,5. *Paired doublets* are very powerful when you find them. In this case there are two incomplete elements in a given row, column or region and both must be one of the two same numbers. Suppose for example that in a given row we have 6,7,8 and 9 already. Then the remaining elements must be 1,2,3,4 or 5, i.e. a linked quintuplet (see Glossary). Now suppose that there is some information which allows us to identify 4,5 as a paired doublet, then the original quintuplet has been reduced to a triplet 1,2,3 and a doublet 4,5 - quite a bit of progress.

Also, in much of what follows, the text refers to numbered regions, as indicated by the numbers in the centre of the regions in the otherwise empty Sudoku grid in Figure 1.1. Hence, the top left hand region is here referred to as region 1, the bottom right as region 9 etc.

2.2 Solution of a Moderate Puzzle (3*)

							2	5
							7	
8								
6						1		3
			7		4			
			2					
	5	3	1					
	2				5			
			6			3		

Figure 2.1

We start with the above Sudoku puzzle, which turns out not to be too bad. Whereas in my earlier book I recommended starting from one region, say top left, region 1, and moving systematically through the grid, when the puzzle is as sparse with poles as these, a different strategy is a lot faster. In particular, the first thing to look for is a region which contains three poles arranged in a triangle as in regions 3 and 7. In region 7, we note immediately that 6 is excluded from three elements in row 9 and 2 in column 1. This means that we can place the 6 in row 8, column 3. Also, in region 3, the triplet (1,3,4) is excluded from column 7 and this must therefore leave a linked triplet in the remaining three elements in columns 8 and 9 of region 3. The result is the following, where we have also completed column 7 with the linked triplet (6,8,9) in region 3 and the linked triplet (2,5,7), extending over regions 6 and 9.

						689	**2**	**5**
						689	**7**	14
8						689	3	14
6						**1**		**3**
				7		**4**		
				2		257		
	5	**3**	**1**			257		
	2	6			**5**	257		
			6			**3**		

Figure 2.2

Information in rows 6-8 allows the (2,5,7) triplet in column 7 to be resolved. The position of the 2 and 7 can also be identified in region 6. The remaining elements form linked triplets in regions 6 and 9. Some of these can be further simplified, and in particular the (1,4,5) triplet in region 9 can be resolved, as shown in Figure 2.3.

						689	**2**	**5**
						689	**7**	14
8						69	3	14
6						**1**	89	**3**
			7			**4**	689	2
			2			5	689	7
	5	**3**	**1**			2	4	6
	2	6			**5**	7	1	89
			6			**3**	5	89

Figure 2.3

Quite a few elements have already been resolved in rows 7 and 8 and the possibilities for the remaining elements are considered next. Row 7 contains a linked triplet (7,8,9) which can be simplified slightly. Row 8 contains a linked quadruplet (3,4,8,9), which again can be simplified slightly.

From here the best strategy may well be to work around the grid and to populate the elements as far as possible. It is generally not worth considering quintuplets and above, but quadruplets and simpler linked elements can often be resolved or simplified.

In region 4, the 2 can be identified, which enables the 2 to be placed in regions 8, 2 and 1. This is a common strategy for the solution of sudoku puzzles: to start with a particular number and to see whether it can be placed in as many regions as possible before moving on to the next number. It is not generally a strategy I recommend, but for puzzles with few clues, such as here, this technique can be quite effective and may be tried, as here, after a few elements have been identified.

In region 5, we also see that the (1,6) pair are excluded from all but 2 of the elements, as shown next.

						689	**2**	**5**
2						689	**7**	14
8			2			69	3	14
6		2				**1**	89	**3**
			7	16		**4**	689	2
			2	16		5	689	7
79	**5**	**3**	**1**	89	789	2	4	6
49	**2**	6	3489	3489	**5**	7	1	89
			6		2	**3**	5	89

Figure 2.4

We now exploit one of the common features of Sudoku-17. Note that, in region 5, 3 is excluded from row 4. Thus 3 must be present in column 4, either row 5 or 6, it doesn't matter. This implies that column 4, row 8 can't be 3 and hence the 3 in region 8 can be identified. This leaves a linked triplet (4,8,9) in region 8 so that the 7 can also be identified. Row 7 can now be resolved, leaving a (4,9) paired doublet in region 8 as shown in Figure 2.5.

Row 8 can also be resolved now, as well as region 9. The remaining elements of region 7 form a link triplet (1,7,8) which can be simplified slightly. Column 1 now has 5 identified elements and the remaining elements form a linked quadruplet (1,3,5,7), which can be simplified slightly as shown in Figure 2.6.

						689	**2**	**5**
2						689	**7**	14
8			2			69	3	14
6		2				**1**	89	**3**
				7	16	**4**	689	2
				2	16	5	689	7
9	**5**	**3**	**1**	8	7	2	4	6
49	**2**	6	49	3	**5**	7	1	89
			6	49	2	**3**	5	89

Figure 2.5

137						689	**2**	**5**
2						689	**7**	14
8			2			69	3	14
6		2				**1**	89	**3**
135				**7**	16	**4**	689	2
13				**2**	16	5	689	7
9	**5**	**3**	**1**	8	7	2	4	6
4	**2**	6	9	3	**5**	7	1	8
17	178	178	**6**	4	2	**3**	5	9

Figure 2.6

The 5 can be immediately identified in column 1, leaving a (1,3,7) linked triplet. The 7 can also be identified in region 2, which reduces the (1,7,8) linked triplet in row 9 to a 7 plus a paired doublet (1,8). In regions 2 and 5, we can now write in the linked quadruplets (3,4,5,8) in column 4, (1,5,6,9) in column 5 and (3,4,8,9) in column 6. These elements can be simplified as shown in Figure 2.7.

13			7	169	3489	689	2	5
2			3458	1569	3489	689	7	14
8			2	1569	49	69	3	14
6		2	458	59	489	1	89	3
5			38	7	16	4	689	2
13			348	2	16	5	689	7
9	5	3	1	8	7	2	4	6
4	2	6	9	3	5	7	1	8
7	18	18	6	4	2	3	5	9

Figure 2.7

In region 5, we note that the 9 must be in row 4, column 5 or 6 it doesn't matter. Hence, the 9 cannot be present in row 4, column 8. The identifies the 8 in row 4, column 8, and reduces the elements in region 5, row 4 to a linked triplet (4,5,9). The remaining element in row 4 must be 7 in column 2.

The grid is sufficiently well covered now that we can use standard sudoku techniques starting systematically from region 1 where the 7 can be identified followed by the 5. The 5 can now be identified in region 2, which resolves the (4,5,9) linked triplet in row 4, as shown in Figure 2.8.

13			7	169	3489	689	**2**	**5**
2		5	348	169	3489	689	**7**	14
8		7	2	5	49	69	3	14
6	7	2	5	9	4	**1**	8	**3**
5			38	**7**	16	**4**	69	2
13			348	**2**	16	5	69	7
9	**5**	**3**	**1**	8	7	2	4	6
4	**2**	6	9	3	**5**	7	1	8
7	18	18	**6**	4	2	**3**	5	9

Figure 2.8

Region 2 can now be further simplified with the identification of the 4 and 9 and the reduction of the remaining unknown elements to a (1,6) and (3,8) paired doublets. Row 3 can also be completed, leaving a (8,9) paired doublet in region 3. The (1,4) paired doublet in region 3 can be resolved, followed by the (1,6) paired doublet in region 2, as shown in figure 2.9. From here, the puzzle can be completed very quickly using normal sudoku methods and the solution is shown in Figure 2.10.

2.3 General Remarks

To summarise, solving this puzzle was more complex than many Sudoku puzzles. Although one singlet could be identified immediately, the density of poles limited the progress that could be made in this direction. The main method of starting

Sudoku Seventeen, by John Austin

13			7	1	38	89	2	5
2		5	4	6	38	89	7	1
8		7	2	5	9	6	3	4
6	7	2	5	9	4	1	8	3
5			38	7	16	4	69	2
13			38	2	16	5	69	7
9	5	3	1	8	7	2	4	6
4	2	6	9	3	5	7	1	8
7	18	18	6	4	2	3	5	9

Figure 2.9

3	6	4	7	1	8	9	2	5
2	9	5	4	6	3	8	7	1
8	1	7	2	5	9	6	3	4
6	7	2	5	9	4	1	8	3
5	3	9	8	7	1	4	6	2
1	4	8	3	2	6	5	9	7
9	5	3	1	8	7	2	4	6
4	2	6	9	3	5	7	1	8
7	8	1	6	4	2	3	5	9

Figure 2.10

was to find several examples of paired doublets within regions where the poles formed a triangle. By working between regions several linked triplets could be identified within these regions and this procedure allowed some rows and columns to be partially completed. Once 5 elements were known within a particular row, column or region, further information on the remaining 4 could be deduced.

Another approach that was useful was to use the presence for example of a 3 along a row or column to deduce that within a certain region, the 3 must be present along a particular column or row. In other words information about a row gave iinformation about a column and vice versa. This turns out to be a particularly useful facility in sparsely populated sudoku puzzles.

Finally, after a few elements have been definitely identified it is worth going over the whole grid to see whether the identified numbers can be determined in other regions. It is often possible to complete all 9 examples of a particular number. So starting with 17 known elements if a few can be quickly identified, say raising the number to 20 and then one of the three extra can be put in the remaining regions, you could within a minute or two of starting the puzzle have transformed it from 17 to 26 known elements, similar to a regular Sudoku puzzle that you see published in newspapers and magazines.

This puzzle was fairly straightforward and can be solved in 20 min or so but other puzzles give up their secrets less readily! In the next section we will explore a puzzle which may well take between 30 and 45 minutes to solve.

2.4 Solution of a Difficult Puzzle (4*)

Figure 2.11 is another Sudoku-17 puzzle, but which is more difficult than the previous one. This one took me about 38 min to solve, so would qualify as 4*. For this puzzle, none of the regions contains three elements, so we need to apply a slightly different strategy.

Sudoku Seventeen, by John Austin

							2	4
	1				8			
				7				
6			2		1	5		
4								3
	7					8	1	
5			4	3				

Figure 2.11

Starting in region 9 we see that the pair (3,4) is excluded from all but two elements, and hence the possible values can be written in. Likewise, (3,4) is excluded from the 7th row except in region 7 where the pair can be resolved. Row 7 now has 5 elements in place and the remaining 4 form a linked quadruplet (2,5,6,9) which can be simplified slightly. We now see that the 4s in all 9 regions can now be placed completing them in the order region 1,2,5,6,9. In particular, the 4 now present in column 8 resolves the (3,4) pair in row 9. We have already completed 8 elements, as shown in Figure 2.12.

In region 2, the 2 can be identified, allowing row 7 to be simplified slightly, and the (3,7,8,9) quadruplet in row 4 can be written in. The 3 can also be identified in region 5, and then region 2. This implies that in region 1, the 3 must be present in column 3 (rows 2 or 3). The position of the 3 in region 4 can now be established. Column 6 is now complete except for a linked quadruplet (5,6,7,9) as shown in Figure 2.13.

25

							2	4
	1			4	8			
	4			7				
6			2		1	5	4	
4								3
					4			
3	7	4	569	2569	2569	8	1	2569
5			4	3				
						4	3	

Figure 2.12

					3		2	4
	1			4	8			
	4			7	2			
6	3	789	2	89	1	5	4	789
4					5679			3
				3	4			
3	7	4	569	2569	569	8	1	2569
5			4	3	679			
					5679	4	3	

Figure 2.13

26

Sudoku Seventeen, by John Austin

The unknown elements of region 2 form a quadruplet (1,5,6,9), linked within the region. Three of these elements occur in column 4 and with the element in row 7 form the same quadruplet but with the linking along the column. The remaining elements of column 4 therefore form a paired doublet (7,8). This is a common feature of Sudoku-17 puzzles: what would otherwise have been a difficult-to-resolve, linked sextuplet along column 4 has been separated into a linked quadruplet and paired doublet with hardly any effort.

			1569	569	3		2	4
	1		569	4	8			
	4		1569	7	2			
6	3	789	2	89	1	5	4	789
4			7	5689	569			3
			3	5689	4			
3	7	4	569	2	569	8	1	569
5			4	3	679			
			8	1	5679	4	3	

Figure 2.14

We can now identify the 1 in region 8 and the (7,8) paired doublet can be resolved. Also, in region 8, the position of the 2 is forced. Further simplifications are possible in row 7, which now contains a (5,6,9) linked triplet. The remaining elements of column 5 now from a linked quadruplet (5,6,8,9) as shown in Figure 2.14.

Sudoku Seventeen, by John Austin

789	569		1569	569	3		2	4
279	1		569	4	8			
89	4		1569	7	2			
6	3	789	2	89	1	5	4	789
4	259		7	5689	569			3
1	259		3	5689	4			
3	7	4	569	2	569	8	1	569
5	8	1	4	3	679	2679	679	2679
29	269	269	8	1	57	4	3	57

Figure 2.15

In column 1, we can identify the position of the 1, leaving a linked quadruplet (2,7,8,9) which can be simplified. Another thing to notice is that in row 9, the pair (5,7) is excluded from all but two elements and so (5,7) must be present in columns 6 and 9. The other unknown elements of row 9 therefore form a linked triplet (2,6,9). The other unknown elements of region 7 form a paired doublet (1,8) which can be resolved.

It is now worth populating the remainder of row 8, which requires a linked quadruplet (2,6,7,9), and column 2, which requires a linked quadruplet (2,5,6,9). Each linked quadruplet simplifies slightly as shown in Figure 2.15.

In Row 1, the 8 must be in column 1 or 3. If it is in column 3, it is excluded from the whole of column 1. Therefore row 1, column 1 must be 8. Column 1 can now be resolved, and the (2,6,9) linked triplet in row 9 reduces to a paired doublet (6,9). The remaining unknowns in region 1 form a linked quadruplet (2,3,5,6), which immediately separates into two paired doublets (5,6) and

(2,3). The (2,3) pair can be resolved and most of the unknown elements in row 1 can also be determined. Both regions 1 and 2 are now complete except for the presence of (5,6) paired doublets. The 9 in column 4 can be established, leaving a (5,6) paired doublet in row 7, as shown in Figure 2.16.

8	56	56	1	9	3	7	2	4
7	1	2	56	4	8			
9	4	3	56	7	2			
6	3	789	2	89	1	5	4	789
4	259		7	5689	569			3
1	259		3	5689	4			
3	7	4	9	2	56	8	1	56
5	8	1	4	3	679	269	679	2679
2	69	69	8	1	57	4	3	57

Figure 2.16

The 9 in column 5 enables the remainder of the column to be simplified. The 9 is also established in region 5 and rows 4 and 5 can be further simplified.

We can continue from region 3, where the 3 can be immediately identified. The other unknown elements can be written in as a linked triplet (5,6,9) along row 2 and a linked quadruplet (1,5,6,8) along row 3. For completeness (while we await inspiration!) we fill in the remaining elements of regions 4 and 6. While proceeding with this, we see that column 7, row 5 is 1. This allows further simplifications to be made in regions 2 and 3 as shown in Figure 2.17.

Sudoku Seventeen, by John Austin

8	56	56	1	9	3	7	**2**	**4**
7	**1**	2	6	4	**8**	3	59	59
9	4	3	5	**7**	2	6	8	1
6	3	79	**2**	8	**1**	**5**	4	79
4	25	58	7	56	9	1		**3**
1	259	5789	3	56	4	29		
3	**7**	4	9	2	56	**8**	**1**	56
5	8	1	**4**	**3**	679	29	679	2679
2	69	69	8	1	57	4	3	57

Figure 2.17

8	56	56	1	9	3	7	**2**	**4**
7	**1**	2	6	4	**8**	3	5	9
9	4	3	5	**7**	2	6	8	1
6	3	79	**2**	8	**1**	**5**	4	79
4	25	58	7	56	9	1	69	**3**
1	259	5789	3	56	4	29	679	2678
3	**7**	4	9	2	56	**8**	**1**	56
5	8	1	**4**	**3**	679	29	679	267
2	69	69	8	1	57	4	3	57

Figure 2.18

Sudoku Seventeen, by John Austin

We now note that in region 9, the 5 must be in column 9, row 7 or 9. This implies that row 2, column 9 can't also be 5, so it must be 9. Region 3 can now be completed and the possibilities for the remaining elements of region 6 can be written in. We have reached the position shown in Figure 2.18. From here the final steps are trivial and the solution is shown in Figure 2.19.

8	6	5	1	9	3	7	**2**	**4**
7	**1**	2	6	4	**8**	3	5	9
9	4	3	5	**7**	2	6	8	1
6	3	9	**2**	8	**1**	**5**	4	7
4	2	8	7	5	9	1	6	**3**
1	5	7	3	6	4	2	9	8
3	**7**	4	9	2	5	**8**	**1**	6
5	8	1	**4**	**3**	6	9	7	2
2	9	6	8	1	7	4	3	5

Figure 2.19

This puzzle was more difficult to solve than the puzzle in Chapter 2.2, primarily because of more complex interactions which needed to be resolved. Similar techniques were used but the extra interactions gave rise to a more complex grid (e.g. compare Fig. 2.16 with Fig. 2.7) before the final resolution of the elements.

Puzzles of 3 and 4* complexity constitute just under 75% of Sudoku-17 puzzles, here represented by puzzles 1-150. In the next section, we get serious!

2.5 Solution of a Very difficult/Atomic Puzzle (5/6*)

			6	8				
1								
						2		
4	3							1
			2			7		
			9			3		
			1	4				
		2		9				
	5			3				

Figure 2.20

The above puzzle took me 59 min, so it is on the border between my two difficulty classes. I have chosen it because, apart from the usual Sudoku-17 ideas, it has an interesting twist just as it reaches its maximum complexity. In other words, in solving a sudoku puzzle, the easy elements are determined first and it becomes gradually more difficult to make further progress until a peak difficulty is reached. Once you get over that peak, the puzzle becomes easier again. If you feel the need to use trial and error, the maximum complexity should coincide with the start of the trial and error process.

I have often wondered whether you can determine the complexity of a puzzle by inspection and perhaps a few brief calculations, but I have so far failed to arrive at anything solid. For example, you might think that having very few poles in each row, column and region would make things more difficult, but in

Sudoku Seventeen, by John Austin

Sudoku-17 the total number of poles is constant so if some regions are more sparse, others are more populated. Perhaps then, an indication of complexity is given by the variance of these numbers. For example a puzzle with many poles in one region and none in several regions might be very difficult. However, it might also be easier as by concentrating on the extra data region, the solution might proceed more quickly. I am still thinking of these ideas and whether you can determine the complexity of sparse puzzles by "inspection", but in the meantime we proceed with the given puzzle.

To start this puzzle, we can try out a few ideas until one or two get us started. For example, we see that in region 9, the 3 is excluded from all elements except two in column 9. Likewise the 1 in region 7 must be confined to one of just two elements. Unfortunately this doesn't help us much. So, what we are trying to do is use two of the same number: one in a row and one in a column to provide useful information.

After a little effort, we see that row 5, column 5 is 3. In region 5, the 1 is in column 6, either row 5 or 6. This implies that in region 2, the 1 is eliminated from columns 5 and 6 and hence it is identified in row 3 column 4. The 3 in region 2 can now also be determined. In column 4, we now write in the linked quadruplet (4,5,7,8), which simplifies slightly.

We can use a similar method for 2 as we did for 1. In region 4, the 2 must be in row 6, which allows the 2 to be determined in region 6. Now we can determine the 4 and 9 in region 6. The region 6 is complete except for a (5,6,8) linked triplet in column 7. the remaining unknown elements of column 7 form a linked triplet (1,3,7), which can be resolved, as shown in Figure 2.21.

We now note that regions 3, 6 and 9 as sets of linked triplets (1,4,9), (2,3,7), (5,6,8) and hence the possibilities for the elements of these regions can be written in. The triplets can be further simplified, as shown in Figure 2.22.

Sudoku Seventeen, by John Austin

			6	8		3		
1			3			7		
			1			2		
4	3		578			568	2	1
			2	3		568	7	9
			9			568	3	4
			578	1		4		
	2		4578			9		
	5		478		3	1		

Figure 2.21

			6	8		3	1	5
1			3			7	49	68
			1			2	49	68
4	3		578			568	2	1
			2	3		568	7	9
			9			568	3	4
			578	1		4	568	237
	2		4578			9	568	37
	5		478		3	1	68	27

Figure 2.22

Sudoku Seventeen, by John Austin

279	2479	47	6	8	279	3	1	5
1			3	2459	259	7	49	68
			1	4579	579	2	49	68
4	3	9	578	567	5678	568	2	1
			2	3	4	568	7	9
			9	567	1	568	3	4
			578	1		4	568	237
	1	2	4578			9	568	37
	5	4	78		3	1	68	27

Figure 2.23

The linked quadruplet on Row 1 can be written in while we await inspiration! We also write in the linked quintuplet (2,4,5,7,9) in region 2. Note also that the 9 in region 4 must be in row 4 column 3, and hence the (5,6,7,8) linked quadruplet can be written in row 4.

In region 7, 1 and 4 and be placed. The 4 can be placed in region 5, which allows the elements in region 2, column 6 to be simplified slightly. The 1 can now be placed in region 5, which leaves the remaining element as a member of the linked quadruplet (5,6,7,8), as shown in Figure 2.23.

The 4 can be placed in column 4, leaving a linked triplet (5,7,8) along column 4. In region 8, the possibilities for the two unknown elements on row 8 can be written in. We can now see that there is a linked triplet (5,6,7) along column 5 which allows us to simplify further the elements in column 5. In particular, row 9, column 5 is seen to be an element of the linked triplet (2,4,9). Region 8 is now also seen to contain a linked quadruplet (5,6,7,8) so the remaining unknown element, at row 7, column 6 must be 2 or 9. The 6 unknown elements of column 6 are now seen to reduce to a linked quadruplet (2,5,7,9) and a paired doublet (6,8).

In region 1, row 1 can be simplified considerably, leaving a (2,9) paired doublet along the row. The position of the 1 can now be identified in column 3, leaving a linked triplet (5,6,8) along row 5. The remaining unknown elements of region 4 can be completed as members of a linked quintuplet (2,5,6,7,8), as shown in Figure 2.24.

29	4	7	6	8	29	3	1	5
1			3	249	259	7	49	68
			1	49	579	2	49	68
4	3	9	578	567	68	568	2	1
568	68	1	2	3	4	568	7	9
256 78	2678	568	9	567	1	568	3	4
			578	1	29	4	568	237
	1	2	4	567	68	9	568	37
	5	4	78	29	3	1	68	27

Figure 2.24

Further progress from here is challenging, and so we systematically write in the possibilities for the remaining unknown elements, starting from region 1.

Region 4 can be simplified as the quintuplet separates into a linked triplet (5,6,8) and a paired doublet (2,7). Row 6 can be tidied up. The results are shown in Figure 2.25.

Sudoku Seventeen, by John Austin

29	4	7	6	8	29	3	1	5
1	2689	568	3	249	259	7	49	68
35689	5689	3568	1	49	579	2	49	68
4	3	9	578	567	68	568	2	1
568	68	1	2	3	4	568	7	9
27	27	568	9	56	1	568	3	4
36789	36789	368	578	1	29	4	568	237
36789	1	2	4	567	68	9	568	37
6789	5	4	78	29	3	1	68	27

Figure 2.25

29	4	7	6	8	29	3	1	5
1	29	68	3	249	5	7	49	68
35	68	35	1	49	7	2	49	68
4	3	9	578	567	68	56	2	1
568	68	1	2	3	4	568	7	9
27	27	568	9	56	1	568	3	4
36789	79	368	578	1	29	4	568	237
3678	1	2	4	567	68	9	568	37
6789	5	4	78	29	3	1	68	27

Figure 2.26

Sudoku Seventeen, by John Austin

In row 3, the 9 is included in the (4,9) paired doublet, so can be excluded from columns 1,2 and 6. The only position for the 7 in row 3 is seen to be in column 6. The 5 in region 2 must also be in column 6. Also in row 3, column 2 cannot be 5, which is already present in column 2 (row 9). Hence the (3,5,6,8) quadruplet in row 3 can be reduced to two paired doublets (3,5) and (6,8). The remaining unknown elements of region 1 can be reduced to two paired doublets (2,9) and (6,8).

In region 5, the 8 must be in the 4th row. Thus, 8 is excluded from row 4, column 7. The element in row 7, column 2 can also be considerably reduced since 3 is already present in the column, and there is also now a (6,8) paired doublet in the column. The results are shown in Figure 2.26, and from here, there are no obvious ways of progressing.

It took me about 45 min to reach this point in solving the puzzle and I then reverted to trial and error methods, starting with row 1. I tend to think that the best strategy is to find options that prove to be incorrect as it tends to speed up the process, especially if like me, you are keen to demonstrate that the solution is unique.

In row 1, the 2 must be in column 1 or 6. If the 2 is in column 6, we have an interesting situation alluded to at the beginning of this section. The 2 in region 2 would be eliminated from row 2, column 5 and the paired doublet (4,9) would be present in both column 5 and 8. A moment's thought, however, would suggest that this would mean that the linked pair could not be resolved. To resolve the pair, we need a 4 or 9 either from another element on the column or another point on the row. In both cases this is not possible. So, if the solution exists, it is not unique. Of course this doesn't represent proof that the solution doesn't exist but it is a hint. Let us explore setting the element row 1, column 6 to 2 and run through the possibilities. We quickly arrive at Figure 2.27.

We see that there is a clash of 2s along column 6 (indicated by the asterisk). It follows that, as surmised, row 1,

Sudoku Seventeen, by John Austin

9	4	7	6	8	2	3	1	5
1	29	68	3	49	5	7	49	68
35	68	35	1	49	7	2	49	68
4	3	9	578	567	68	56	2	1
568	68	1	2	3	4	568	7	9
27	27	568	9	56	1	568	3	4
3678	9	368	578	1	2*	4	568	237
3678	1	2	4	567	68	9	568	37
678	5	4	78	29	3	1	68	27

Figure 2.27

2	4	7	6	8	9	3	1	5
1	9	68	3	2	5	7	4	68
35	68	35	1	4	7	2	9	68
4	3	9	578	567	68	56	2	1
568	68	1	2	3	4	568	7	9
27	27	568	9	56	1	568	3	4
36789	79	368	578	1	2	4	568	37
3678	1	2	4	567	68	9	568	37
678	5	4	78	9	3	1	68	2

Figure 2.28

2	4	7	**6**	**8**	9	3	1	5
1	9	8	3	2	5	7	4	6
5	6	3	1	4	7	**2**	9	8
4	**3**	9	5	7	8	6	2	**1**
6	8	1	**2**	3	4	5	**7**	9
7	2	5	**9**	6	1	8	**3**	4
9	7	6	8	**1**	2	**4**	5	3
3	1	**2**	4	5	6	**9**	8	7
8	**5**	4	7	9	**3**	1	6	2

Figure 2.29

column 6 cannot be 2, and therefore must be 9.

Figure 2.28 shows an intermediate stage during the evaluation of the elements, and continuing this procedure leads to the final solution in Figure 2.29.

Overall, it was quite challenging to unravel all the complex interactions. My method of solution is somewhat *ad hoc* but is faster than working systematically from region to region as it concentrates on different possibilities as they arise. This of course requires a little experience but is not really that difficult. The Sudoku-17 puzzle solved here was not the most difficult of the 5/6* classes in the puzzle section. Some of these require several attempts with trial-and-error methods in the final stages. This can of course be quite time consumiing. I wish you the best of luck!

2.6 The World's Most Difficult Sudoku Puzzle

If, after solving all the puzzles, you find them easy enough, try the following one, but do this last. It is reproduced again as the first one in the puzzle section. It contains a few more than the minimum number of 17 poles (in fact it has 21), but it was composed by the Finish mathematician Arto Inkala and has been rated by some as the world's most difficult Sudoku puzzle[1]. However, another source[2] suggests that the Inkala puzzle is just the third most difficult, in what is a group of a small number of puzzles which might be considered unsolvable by normal methods. So, you are warned! Using a trial and error method, I managed to solve it in a few days and some brief comments are given before the solution (page 166). Enjoy!

8								
		3	6					
	7			9		2		
	5				7			
			4	5	7			
			1				3	
		1					6	8
		8	5				1	
	9				4			

Figure 2.30: This is not a Sudoku-17 puzzle, but has been rated by some mathematicians as the world's hardest[1].

Sudoku Seventeen, by John Austin

References

[1] World's hardest Sudoku: can you crack it?, Nick Collins, Daily Telegraph, 28 June 2012, http://www.telegraph.co.uk/science/science-news/9359579/Worlds-hardest-sudoku-can-you-crack-it.html), accessed 10 December 2014.
[2] Arto Inkala Sudoku, Andrew Stuart, http://www.sudokuwiki.org/Arto_Inkala_Sudoku, accessed 10 December 2014.

Glossary

Chain: A set of *elements* joined by a line, which may be multi-branched.

Chain Sudoku: A Sudoku-like puzzle in which the numbers 1-9 need to be present once and only once along each *chain*.

Column: The nine *elements* connected in a single straight line from top to bottom in the Sudoku puzzle.

Condensation: The process of resolving a *doublet* or *triplet*.

Continuity: The method of completion of a *row, column* or *region* when 8 of the 9 *elements* are known.

Doublet: An unfilled *element* of the Sudoku *grid*, which on careful examination can have one of only two values. (see also *paired doublet*)

Element: One of the 81 squares making up the standard Sudoku puzzle.

Gefbadchi: A puzzle mathematically identical to Sudoku, in which the numbers 1-9 are replaced by the letters A-I.

Idoku: A sudoku published by i newspaper, which has the same rules as Sudoku, but in addition requires a shaded region in the form of a lower case i to contain the digits 1-9 only once each.

Killer Sudoku: A puzzle with the same rules as Sudoku, but instead of *poles*, the initial data are supplied in the form of sums of two or more adjacent Sudoku *elements*.

Latin Square: A square of any size containing a set of symbols each of which appears only once in any *row* or *column*.

Linked: Used in the phrase *linked doublet, linked triplet, linked quadruplet* etc. These are groups of unresolved *elements* which occur in the same *row, column* or *region* with the same possible vaues.

Magic Square: A square of any size containing numerical elements which sum to give the same value in each *row, column* and long diagonal.

Pole: An *element* of the Sudoku puzzle which is initially specified.

Sudoku Seventeen, by John Austin

Paired Doublet: Two *elements* in the same *row, column* or *region* which both contain one of only two values. Another term for these *elements* is *linked doublet*.

Populating the Grid: Putting all the numbers possible in every grid element. These numbers are often written in small print or in pencil so that they can be erased or covered up once the element is resolved.

Related: Two *elements* are related if they occur in the same *row, column* or *region*.

Quadruplet: An unfilled *element* of the Sudoku *grid*, which on careful examination can have one of only four values. (See also *linked*)

Quintuplet: An unfilled *element* of the Sudoku *grid*, which on careful examination can have one of only five values. (See also *linked*)

Region: The nine *elements* arranged in a 3x3 square in the sudoku puzzle confined to *rows* 1-3, 4-6 or 7-9, and *columns* 1-3, 4-6 or 7-9.

Row: The nine *elements* connected in a single straight line from left to right in the Sudoku puzzle.

Reduction: Partial solution of a Sudoku *element*, resulting in the number of possible values for the *element* being reduced in number from for example 4 to 3 or 3 to 2.

Singlet: An unfilled *element* of the Sudoku *grid*, which on careful examination can have only one value.

Starting Element: The sudoku grid *element* that provides the start of an attempt to complete the puzzle by trial and error.

Sudokarrow: A Sudoku-like puzzle which combines as initial starting information a set of *poles* together with a set of paths (arrows) along which the *elements* sum to the value at the start of the path.

Sudoku: A puzzle typically consisting of a 9x9 grid containing specified *elements* (*poles*) with the objective of completing the empty squares of the grid such that each *row, column* and *region* contains the numbers 1-9.

Sudoku Seventeen, by John Austin

Tredoku: A form of three-dimensional Sudoku, based on sides of a cube and beloved of some *Times* readers.

Trial and error: The substitution of plausible values for unresolved *elements* until the puzzle is solved. Since the method often leads to error, backtracking through the puzzle is needed.

Triplet: An unfilled element of the Sudoku *grid*, which on careful examination can have one of only three values. (See also *linked*)

Sudoku Seventeen, by John Austin
Acknowledgements

I am grateful to Gordon Royle of the University of Western Australia for making his collection of sudoku-17 puzzles available. This print book was prepared with openoffice software and supplied to CompletelyNovel.com for the production of print on demand copies.

Sudoku Seventeen, by John Austin

John pictured at Bryce National Park, Utah, USA

About the Author

The Author, Dr. John Austin, has over 30 years' research experience on the upper atmosphere and has published over 80 papers in numerous international scientific journals. In addition John worked for 4 years as an Editor of the Journal of Geophysical Research, the premier Geophysics journal in the USA.

He has spent several years working in the USA, at NASA Langley, Hampton, virginia (1984-1985) and the University of Washington (1988-1990), where amongst other things he met his future wife Alda, to whom he is still married. During 2003-2011 John worked in Princeton, NJ, USA. His main scientific contribution has been to show the connection between ozone depletion and climate change. John has been involved in the writing of numerous international reports for the World Meteorological Organisation

47

Sudoku Seventeen, by John Austin

and The Intergovernmental Panel on Climate Change, for which the IPCC received the 2007 Nobel peace prize.

In recent years, John has broadened his work into popular science, through the website DecodedScience.com and in 2014 he created an internet scientific publishing business Enigma Scientific Publishing, http://www.enigmascientific.com. "Measuring the World" was his first popular science book, available in hardcopy and as an ebook.

****

When not working, John enjoys a variety of activities including chess, running, photography and travel. He has become addicted to sudoku, which is why this book has been written!

The Puzzles

Do This LAST!

8								
		3	6					
	7			9		2		
	5				7			
			4	5		7		
		1				3		
	1						6	8
	8	5				1		
	9				4			

This is not a Sudoku-17 puzzle, but has been rated by mathematicians as the world's hardest (http://www.telegraph.co.uk/science/science-news/9359579/Worlds-hardest-sudoku-can-you-crack-it.html).

Target: 25 min(est.)

001

						1		
			5					
2								
	3	5	4					
				1		8		
	7					2		
		4					9	5
6				2				
					8		3	

Target: 25 min(est.)

002

						1	3	5
8		7						
							4	
	3		4				1	
				7	6			
	5							
2						7		6
						2		
			3					

51

003

Target: 25 min(est.)

004

Target: 25 min(est.)

Sudoku Seventeen, by John Austin

Target: 25 min(est.)

005

					4	1		
	2							
	9							
4							3	7
1			5					
							2	
			2	9		6		
7	4					8		
			3					

Target: 25 min(est.)

006

					8	3		
7			4					
	1							
							1	5
4			6					
						9		
3		8				2		
2						6		
			1	9				

Sudoku Seventeen, by John Austin

007

						6		1
9				2				
	3	6			5			
				8		4		
	1							
5			3				9	
4		7					2	
			1					

008

						7	1	
		5	2					
8				3				
						4		5
							6	2
2			6		4			
1	7					3		
			5					

54

Sudoku Seventeen, by John Austin

Target: 25 min(est.)

009

					1		2	3
	4							
							1	
	8	2					6	
				7		4		
			3					
			9	4		7		
1						5		
2								

Target: 25 min(est.)

010

				1	6			
2						3		
	1	8						7
			4	3		5		
		6						
	7					1		
			2	5				
4			3					

Sudoku Seventeen, by John Austin

011

					4	1		
		2						
	9							
4							3	7
1			5					
							2	
			2	9		6		
7	4					8		
			3					

012

					6			1
7				5				
4						5	8	
5							7	
		2	1					
	1	6				3		
			4	8				
	2							

Target: 25 min(est.)

013

Puzzle 013:

					6	2		7
	1		5					
		3				4		
	5							1
				9	2			
7								
		4	3					
			8	1				
2								

Target: 25 min(est.)

014

Puzzle 014:

					6			1
7				5				
4						5	8	
5							7	
		2	1					
	1	6				3		
			4	8				
	2							

Sudoku Seventeen, by John Austin

Target: 25 min(est.)

015

					9	8		4
2				7				
1								
7						3	2	
	4		1					
						7		
	6		2				8	
			5					
			3					

Target: 25 min(est.)

016

			1			2		
4							6	
5								
			5		4			7
8			6					
						1		9
	2	7						
			4				3	
	1							

58

Sudoku Seventeen, by John Austin

017

			3			7		6
	1		4	2				
6		3	1					
7							8	
						2		
			6					3
	2						4	
5								

018

				3	7		2	
1		8				5		
9								
5			8			1		
	7							6
	6						7	3
	4		1					

Sudoku Seventeen, by John Austin

				4		6		
	8							
	3							
4				5			7	
			2				8	
6		9						
	1		8					
					3			2
5						4		

				4		8		
		2	3					
3								
7	8						1	
			4		2			
			5					
6						2		3
	7			1				
						4		

60

Sudoku Seventeen, by John Austin

				4	1			
	9					3		
				7			4	1
	2		3					
							9	
1			5			8		
7		4						
6			8					

				5			1	
3			6					
2								
	7		3			6		
	1		8					
	4							
7						3		
				1				2
				4	9			

Sudoku Seventeen, by John Austin

023

				5			9	1
2	3		6					
4								
6			4			3		
				1			8	7
	1			7				
			3			2		

024

				5		3		8
9								
						7		
			4		9		2	
7		1						
			2					
					6			4
	5						9	
	3			7				

Sudoku Seventeen, by John Austin

025

				5		8	4	
1				3				
9								
	8		1			5		
	2						7	
			9		6			
6								1
			7	8				

026

				5	6			
				3			8	
1								
	7	5				3		
			8				4	
	6		1					
8			2					7
						5		6

Sudoku Seventeen, by John Austin

Target: 24 min

027

Puzzle 027

					6			1
	2		4					
6	5					3		
				8	1			7
			3		2	4		
1		8						
7			5					

Target: 29 min

028

Puzzle 028

			7	1				8
6		3						
			4			7	1	
2			6					
5								
	4		5			6	2	
	1							
						3		

Sudoku Seventeen, by John Austin

				9			6	2
4	1							
							3	
7						8	5	
			2					
			6					
		6	3					
1						7		
				8		4		

			1		7	3		
9							1	
8			4					
				8				2
	1					4		
				9				
			3			5		
	7			6				
							8	

Sudoku Seventeen, by John Austin

031

			2				6	
			8		1			
4						5		
							3	8
5				4				
								1
			4	7		2		
		3				9		
	1							

032

			3					6
	1			4				
	5							
				7		5	1	
6			2					
3						9		
8			6		2			
					3			
						1		

Sudoku Seventeen, by John Austin

			3			6	7	
	1							
			7					
6						1		5
4			8					
7								
				1	5	2		
3		8					4	

033

			4			2		
6		5						
		8						
1						3		6
	2		7					
			5				4	
				3	8			
	5					1		
				6				

034

Sudoku Seventeen, by John Austin

035

```
. . . | 4 1 . | 5 . .
. 8 3 | . . . | 6 . .
. . . | . 2 . | . . .
------+-------+------
4 . . | . . . | . 2 1
. . . | 7 . 3 | . 8 .
. . . | . . . | . . .
------+-------+------
1 . . | . . . | . . 5
. . . | 6 . . | 3 . .
. . . | . . . | . . .
```

036

```
. . . | 7 . . | 8 4 .
. . 1 | . . . | . . .
9 . . | . . . | . . .
------+-------+------
5 . . | . . . | 2 . 1
. 3 . | 9 . . | . . .
. . . | . . . | . . .
------+-------+------
. 4 . | . . 6 | 3 . .
. . . | . 2 . | 7 . .
. . . | . 5 1 | . . .
```

68

Sudoku Seventeen, by John Austin

037

		1				6		
			3					
				8				
			5	2				1
7					6			
	4						8	
				1	2			
3	9							
8			7					

038

	1					6		
				3				7
7						1	2	
3				8			5	
						4		
5		6						3
			2		1			
			4					

Target: 22 min

039

★ ★ ★

	1			7				
2						8		
				4				
8			2			6		
				1			7	
			8			3		1
7	5		9					
	4							

Target: 27 min

040

★ ★ ★

	1		6	9				
						8		7
			4					
				3	7		9	
8	6							
						5		
7		3	8					
							6	
2								

Sudoku Seventeen, by John Austin

	1	6				4		
			7			5		
	2							
			8		3		2	
3			5					
							6	
7		4		6				
				1				
8								

041

	2			7			4	
		5						6
	4					7	2	
			6		8			
			1					
			5	3		2		
8		6						
1								

042

Sudoku Seventeen, by John Austin

043

	2		8					
	1							
					7			
	3				4		2	
4				7				
				5				
3						5		7
			1			6		
8			2					

044

	2	8						
					1			
	6							
1		3		6		2		
			5			8		
4								
			8	2				
3							1	
			7				4	

72

045

	4			1			6	
								2
							5	
			2		5		3	
	8					4		
			3					
2				8		7		
5		6						
						1		

046

	4		6				3	
	8					1		
						5		7
1		3		5				
							4	
				7				
5						2		
6			4					
			8					

73

047

Target: 22 min

	5			9				
	3						1	
		6					4	
6						8	2	
				5	1			
4								
			7			5		3
			1					
						9		

048

Target: 27 min

		6				8	2	
				7	1			
4								
5							6	
	2		8					
				3				1
			9			2		
1		3						
7								

Sudoku Seventeen, by John Austin

049

Puzzle 049:

	6		2		4			
							1	
			6					
5				1				3
	2					6		
				3				
			7			2		
3	4							
8				5				

050

Puzzle 050:

	7					8	5	
		2			1			
4			6	5				
3								1
						2		
1	3						9	
			7	2				
			8					

Sudoku Seventeen, by John Austin

051

★
★
★

```
. 7 1 | . . . | 8 . .
. . . | 2 . 3 | . . .
. . . | . . . | . . .
------+-------+------
2 . . | . . . | . 3 6
. . . | . . . | . . .
. . . | . . 9 | 5 . .
------+-------+------
. . . | . 1 . | . . .
. 6 . | . 7 . | 1 . .
3 . . | 4 . . | 7 . .
```

052

★
★
★

```
. 8 . | . . . | . . 3
. . . | 4 . . | . 1 .
9 . . | . . . | . . .
------+-------+------
. . 6 | . 5 . | . 4 .
. 1 . | . . 3 | . . .
. . . | . 8 . | 2 . .
------+-------+------
2 . . | 7 . . | 6 . .
. . . | 1 . . | . . .
. . . | . . . | 8 . .
```

Sudoku Seventeen, by John Austin

	9		1		7			
							8	4
			9				6	
	1	7				5		
			4	8				
6		8		3				
						1		
2								

053

1						9		
		2	8					
			3		5			
				1			4	
	3						5	
							6	
	4		5					
8						2		
9						1		

054

Target: 27 min

055

★
★
★

1			6					5
5	3					4		
						6		
				7	3		8	
4								
				2				
		2					7	
	7		4					
			5					

Target: 22 min

056

★
★
★

1		5					4	
			9	3				
4								
2	3			6				
					8		1	5
	2					3		7
				1				
						2		

Sudoku Seventeen, by John Austin

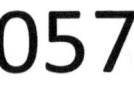

057

2						5	3	
			7	4				
			6				8	
					9	1		
	4							
	6							
5			1		3			
				6				7
9								

058

2		1						
				5		4		
7								
			7	3			2	
	5	4						
							1	
6	7		1					
						5		3
						8		

059

Target: 26 min

2	9					5		
			7			3		
			1		8			
				5		2		
	8						7	
		1						
3				4				
			6				1	
5								

060

Target: 26 min

3			7			4		
2								
1								
	8		2				5	
	4						1	9
			3					6
	6			1				
						2		
			9					

Sudoku Seventeen, by John Austin

061

4						6		9
			7	3				
			5					
2					4	1		
		3						8
						5		
			8				7	3
5		1						

062

4	6		2			5		
			8			1		
						3		
7					6		2	
5				1				
				3				
			5				4	
		1						
	8							

Puzzle 063 — Target: 29 min

5				2	1			
9						4		
				8				
			7				1	
3			4					
							2	
	8			6		3		
				3		9		
	2							

Target: 29 min

063

Puzzle 064 — Target: 27 min

5			6			4		7
			2					
						1		
4		3					9	
				1		8		
	2							
1								5
			3				2	
	8							

Target: 27 min

064

065

Target: 24 min

5		3	6					
					8	1		9
						2		
	9		2					
				5			6	
	1							
8							4	
			1			7		
6								

066

Target: 27 min

5	3						7	
				6	4	2		
	8							
			3				8	7
6				5				
		1						
4						6		
				1				
			8					

067

6								4		
				3	1					
				7					9	3
	1	6								
2										
				5	6			2		
		9			4					
	3							8		

068

6				7		3				
	2							4		
1	4			5						
				2					3	9
	8									
3		9								
					8			5		
					4					

Sudoku Seventeen, by John Austin

Puzzle 069

7						6	5	
			9	3				
1								
8	6		2					
						4		9
			5					
	9	4		8				
							7	
	3							

069

Puzzle 070

7			1					2
	5			3			8	
	4					9	5	
	8		6					
			2					
2		1						
				9		4		
6								

070

85

071

7			6		5			
							8	4
			3			5		
	1	8						
	4			8				
2			5			3		
9		7						
				1				

072

8						6		
7				2				
				1				
			6			2		7
3	5		4					
						1		
4					3			
			8				5	
		2						

Sudoku Seventeen, by John Austin

8			5					9
							1	
4								
	9			1				2
	1			6				
						4		
5			4			7		
						3	6	
			2					

8	3					5		
			1		2		4	
			4					
7	5					6		
			8				1	
	9							
2		1						
				3		9		

Sudoku Seventeen, by John Austin

075

							2	8
1								
							9	
	4		3					
5						7		
			9		8			
			6	5		1		7
				1		3		
	9							

076

	6			8			4	
						7		
	1							
5			2			3		
			1			2		
	8							
				4			6	9
2								8
3								

Sudoku Seventeen, by John Austin

Puzzle 077:

4			3			2		5
			1					
						8		
					7	1		
2				8				
							9	
	3		6					
				2		4		
	6					7		

Puzzle 078:

5						2		
			8		4			
			9					
				1			5	
	8	4						
	6					1		
7			1	2				
							3	4
						5		

Target: 38 min(est.)

079

★
★
★
★

Puzzle 079:

```
. . . | . . . | 2 . 5
1 . . | 8 . . | . . .
. . . | . . . | . . 4
------+-------+------
6 4 . | 5 . . | . . .
. . . | 1 . . | 7 3 .
. 9 . | . . . | . . .
------+-------+------
. . . | . 9 5 | . . .
3 . . | . . . | . 8 .
. . . | . 4 . | . . .
```

Target: 38 min(est.)

080

★
★
★
★

Puzzle 080:

```
. . . | . . . | 4 . 1
7 . . | 2 . . | . . .
9 . . | . . . | . . .
------+-------+------
5 . . | 7 6 . | . 3 .
. 4 . | . . . | 8 . .
. . . | 5 . . | . . .
------+-------+------
. 1 . | . 8 4 | . . .
. . . | . . . | 2 7 .
. . . | . . . | . . .
```

Sudoku Seventeen, by John Austin

Target: 38 min(est.)

081

				1		6		
	4		3					
	8					7		
5			4					2
1		7						
6								
							2	3
				7			8	
			6					

Target: 38 min(est.)

082

				2		7		
	8						5	
		1						
6	4					2		
			1		8			
			5					
2						3		
			4				1	
9			7					

Target: 38 min(est.)

083

★ ★ ★ ★

				2	1			
	3					8		
7		6						2
			3			5		
			9					
			6	5		3		
2	1						7	
4								

Target: 38 min(est.)

084

★ ★ ★ ★

				5			3	8
2			6					
		1						
			1		4	7		
3	5							
	3		2	8				
						1	4	
						5		

Sudoku Seventeen, by John Austin

Target: 38 min(est.)

085

						4	9	
	5		1					
						2		
8			5			3		
3		9						
			2					
				6	9		7	
	2			4				
								1

Target: 38 min(est.)

086

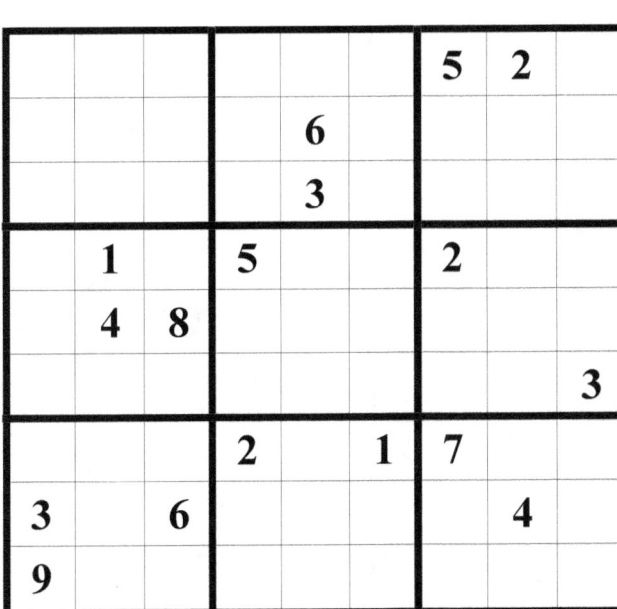

						5	2	
				6				
				3				
	1		5			2		
	4	8						
								3
			2		1	7		
3		6					4	
9								

Sudoku Seventeen, by John Austin

087

★
★★
★
★

					5		6	
	3			2				
8			1		6			
			4			7		3
						2		
	7			3		4		
6			8				5	

088

★
★★
★
★

				1		4		
3							7	
			6					
	4	1		6				
							2	3
						8		
2			3		7			
	8					1		
			5					

94

Target: 38 min(est.)

089

Target: 38 min(est.)

090

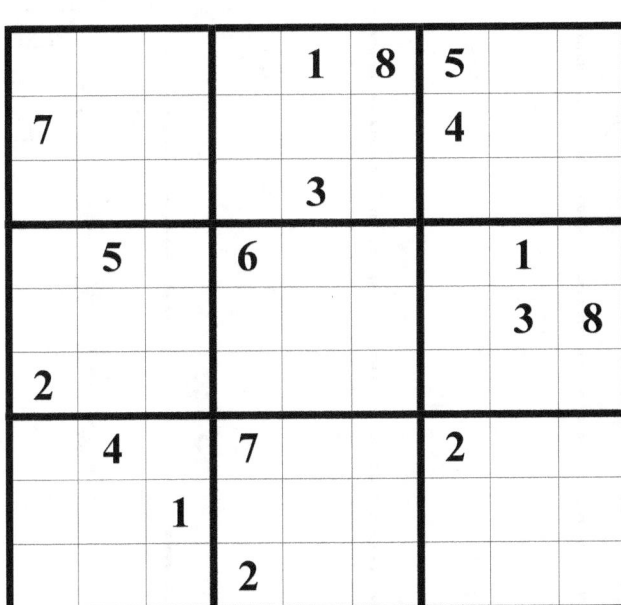

Sudoku Seventeen, by John Austin

091

092

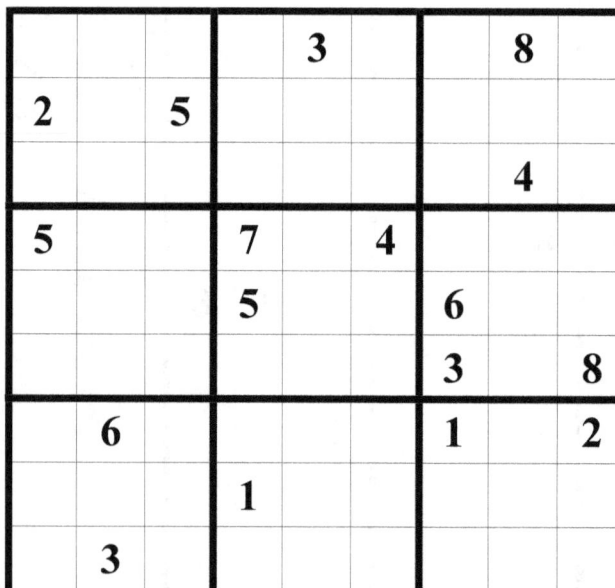

Sudoku Seventeen, by John Austin

				4				1
8			2					
7								
	4			5	1			
2							9	8
		6						
			9			8	7	
	5	3						

Target: 38 min(est.)

093

Target: 38 min(est.)

094

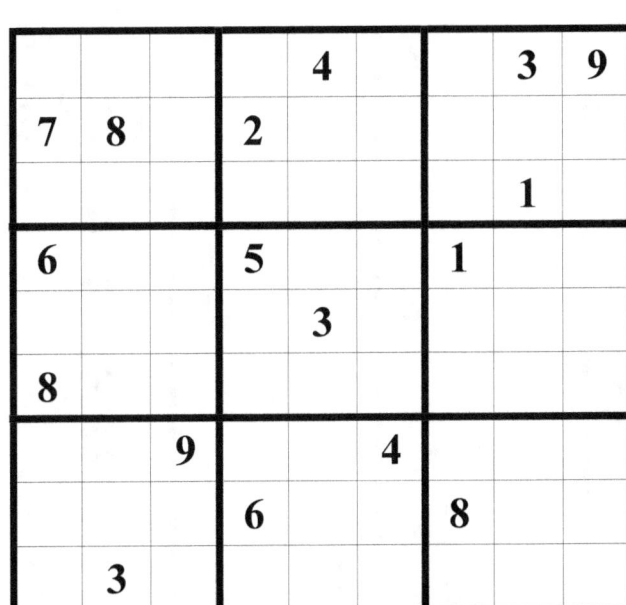

Target: 38 min(est.)

095

Target: 38 min(est.)

096

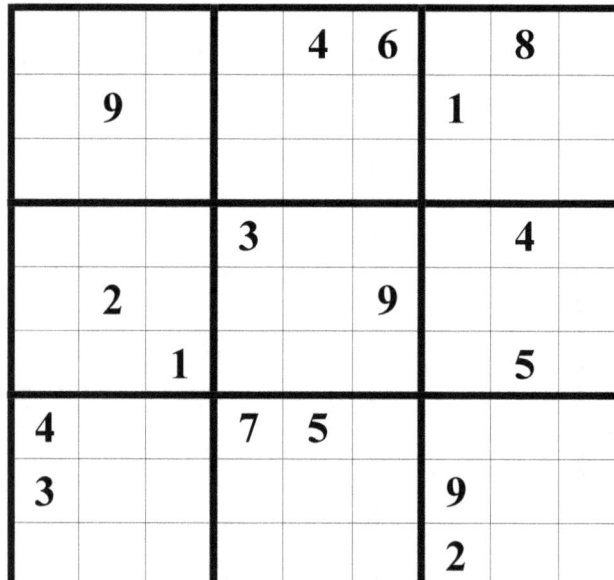

98

Sudoku Seventeen, by John Austin

				5	1		6	
4	2							
	1	6			5			
			3			4		8
		9						
7			2	4				
							1	
						3		

				6			3	1
	2		7					
				5	3			8
	4					2		
				1				
			4			7	2	
5			9					
1								

Sudoku Seventeen, by John Austin

					7		2	
	3		8					
							1	
			3		4	6		
	1							
						3		
7						4		5
5			1	2				
				9				

				8		3	2	
5		6						
4								
			7	3				
7								5
							6	
1	8					7		
			2		6			
				4				

100

Sudoku Seventeen, by John Austin

101

			1			2		
	4					6		
					3			
				8			3	
2				7				
6								
5			6			1		
			2	4				
	8						4	

102

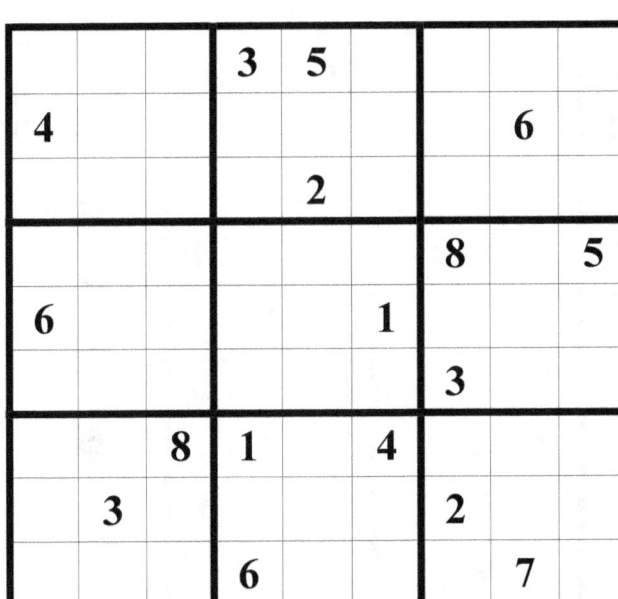

			3	5				
4							6	
				2				
						8		5
6					1			
						3		
		8	1		4			
	3					2		
			6				7	

Sudoku Seventeen, by John Austin

103

			5				8	
2		1						
		7						
4	5		3					
						1		2
						6		
				6	1	2		
	3						4	
				7				

104

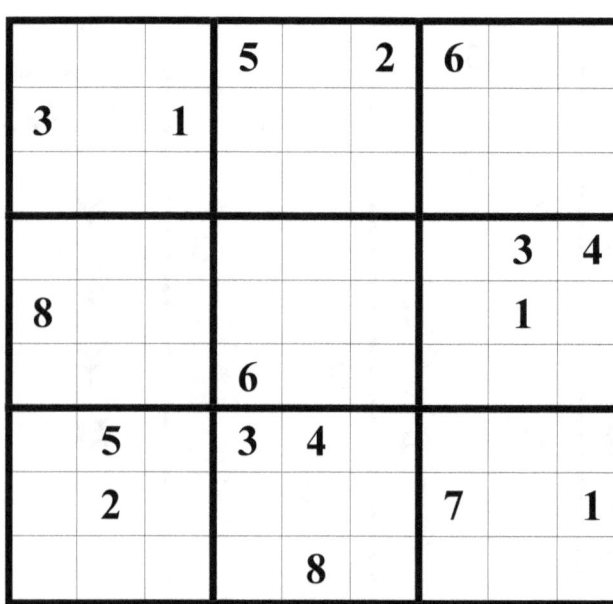

Sudoku Seventeen, by John Austin

105

		6						4
	3	2						
	8							
	5						3	
						9	8	
1								
			8	5	2			
6								7
4			3					

106

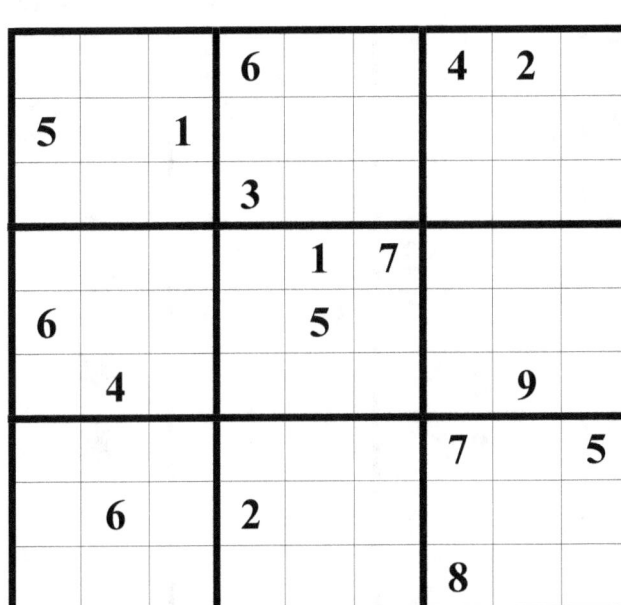

		6			4	2		
5		1						
		3						
			1	7				
6			5					
	4					9		
					7		5	
	6	2						
					8			

Sudoku Seventeen, by John Austin

107

			6		8	1		
4	3							
	7	8				9		
2			3					
			4	5				
		6		9				
							3	2
							4	

108

			9				3	2
9		1	8					
							4	
	3				2			
6						1		
				4				
	2			8				
			7			6		
7								

Sudoku Seventeen, by John Austin

109

★
★
★
★

		2					1	9
				4			5	
			7					
			6			8		
7	4							
	5							
6		8				7		
9					1			
				5				

110

★
★
★
★

		5					7	
	2						6	
				4	1			
			6				2	
4						3		
	5							
3						1		8
		7	5					
						4		

105

Sudoku Seventeen, by John Austin

111

⭐ ⭐ ⭐ ⭐

		6	1					
	2	3						
						8		5
			6	1				
7						4		
			3					
						1	3	
5	7							
4					7			

112

⭐ ⭐ ⭐ ⭐

	2					5		
			1		7			
			3					
1			5					
		9				8		
					4	3		
	6			2				4
3							7	
							1	

Sudoku Seventeen, by John Austin

113

	3		2					
			8	1				
		6				7		
					6			7
1				4				
							2	
						1	6	5
2			3					
						4		

114

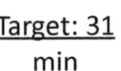

	3	1				5		
			3				6	
		7						
	4					8		
			7			1		
2			6					
6							2	
4				8				
				1				

107

Sudoku Seventeen, by John Austin

Puzzle 115

	4							2
6			1					
						8		
			2	4		5		
3							7	
			9					
			3			1	8	
	2		5					
	9							

115

Puzzle 116

	5		9				8	
			6			3		
1								
6		3	7					
							5	4
9								
	2			5	8			
				6				
				2				

116

Sudoku Seventeen, by John Austin

117

118

Target: 37 min

119

★
★★
★★
★★

	7		3					
6						9		
			1					
5	3	8						
						7	2	
		1						
							1	3
4				9				
				5				7

Target: 31 min

120

★
★★
★★
★★

	8			3	7			
	2					4		
			2			1	8	
3							7	
		6	4					
			6			2		
7				5				
5								

Sudoku Seventeen, by John Austin

121

1							5	
				7			4	
2								
						7	3	
8			4					
			1					
	7	3		4				
			5			1		
	6					2		

122

1			2					
				5		4		
						3		
		6	1				2	
	3					5		
			8					
	4			7				
				3				9
8						1		

Sudoku Seventeen, by John Austin

Target: 31 min

123

★
★
★
★

1	4						3	
			7	2				
	8		6		3			
		6				2		
						1		
		7	5				8	
2			4					
				8				

Target: 32 min

124

★
★
★
★

2							1	
	5		4					
	4					7		2
			3			8		
			1		9			
				5		4		
1			2					
3	6							

Sudoku Seventeen, by John Austin

Target: 43 min

125

2				3	6		7	
								1
5								
				5		3	6	
	4		8					
	1							
3						5		
	8		1					
			4					

Target: 40 min

126

2			4			6		
6							8	
			1					
				8			2	
	7							
		1						
3	9			6				
					5	7		1
						4		

Sudoku Seventeen, by John Austin

Target: 36 min

127 ★ ★ ★ ★

2	1							
			5	8				
6								
	4							8
				2		6		
					1			
	3					2	7	
		5	8					
			4			1		

Target: 35 min

128 ★ ★ ★ ★

3						5		
			8			4		
			9				6	
4				3	1			
	6						8	
	4		7	5				
						3		1
						2		

114

Sudoku Seventeen, by John Austin

3		8		5		6		
			2					7
	2		7			4		
	7		1					
							3	
4				3			8	
								1
6								

129

4								1
			2		5			
			8					
1	3		6					
						2	5	
7								
				4			7	
		8		3				
	2					8		

130

Sudoku Seventeen, by John Austin

4			6			3		7
			2					
						5		
			1					2
3	5							
								9
					3	4	8	
	2			5				
		1						

131

4		7	1					
							6	3
3	6						8	
			7					
	9							
			4	2		7		
5	3				6			
						1		

132

116

Sudoku Seventeen, by John Austin

133

5						2	4	
				8	1			
2								
	1	7	3					
							5	
	8							
	3					7		
			5					8
6			4					

134

5			1			6		7
				3			8	
4								
			2					5
	3					1		
	8							
	9						3	
2			5					
				7				

117

Sudoku Seventeen, by John Austin

Puzzle 135

Target: 31 min

135

★
★
★
★

6								1
				5	2			
			7					
7			3					
3							5	
						2	4	
	5	2				8		
			9	6				
	4							

Puzzle 136

Target: 31 min

136

★
★
★
★

6				3		5		
7							8	
1								
			1		7	4		
	4					3		
			8					
			6				1	
	3			2				
	5							

118

Sudoku Seventeen, by John Austin

6		5		3				
							8		1
	1		2			4		
5				7				6
2								
3							5	
			1		4			
			8					

137

6	3		2					
						5		8
	2					1		
	4						9	
5				8				
				1				
			7				3	
		2	4					
1								

138

119

Target: 43 min

139

★
★
★
★

```
7 .  8 | .  4 . | .  .  .
.  .  . | .  4 . | 1  .  .
.  .  . | .  .  . | .  .  .
------+-------+------
3  6  . | .  1  . | .  .  .
.  .  . | 5  .  . | .  8  7
.  .  . | .  .  . | .  .  .
------+-------+------
.  1  . | .  .  . | 6  4  .
5  .  . | 3  .  7 | .  .  .
.  .  2 | .  .  . | .  .  .
```

Target: 31 min

140

★
★
★
★

```
9  .  . | 5  .  . | .  .  .
.  .  . | .  .  7 | 3  .  .
.  .  . | .  .  . | 1  .  .
------+-------+------
2  .  . | 6  .  . | .  5  .
.  .  . | 3  .  . | .  .  .
.  .  . | .  .  . | .  .  8
------+-------+------
.  .  . | 8  5  . | 2  .  .
.  4  3 | .  .  . | 6  .  .
.  1  . | .  .  . | .  .  .
```

Sudoku Seventeen, by John Austin

141

				6	3	7		
8	1							
			1	5			4	2
3								
			8					
6		3			7			
			4				1	
					2			

142

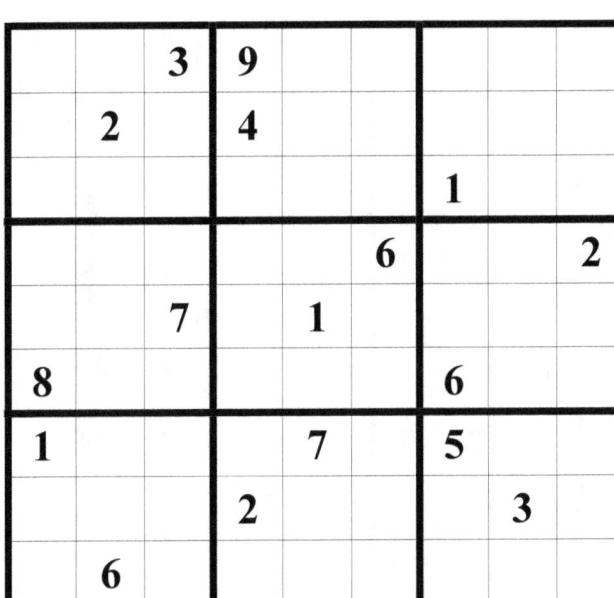

Sudoku Seventeen, by John Austin

6		7			8		2	
1			4					
5							3	
4				5				
				3				
			6					
	3		2					
	9					5		
						1		

143

4	6							7
			5			9		
			2	6			5	
	3							9
6			8					
1				7	3			
						5	2	

144

Sudoku Seventeen, by John Austin

			1			8		6
2			4					
5						3		
					2	7	5	
	6			3				
	1							
			6				2	
3				8				

145

							4	1
2			5					
6								
			6			2	5	
	1		3					
	4					7		
				4	2			
8				7				
						5		

146

123

Sudoku Seventeen, by John Austin

147

3		6					9	
5					1			
7	1		2					
				4			6	
	8							
	4					8		
			9	5				
						3		1

148

	7					5		2
				6		7		
					8			
			7	2		4		
1							6	
3								
	2		5					
6							3	
					1			

Sudoku Seventeen, by John Austin

149

2		6						3
			5	1		4		
8								
				4		6	1	
5	3							
							9	
			7		2			
	1							
			3					

150

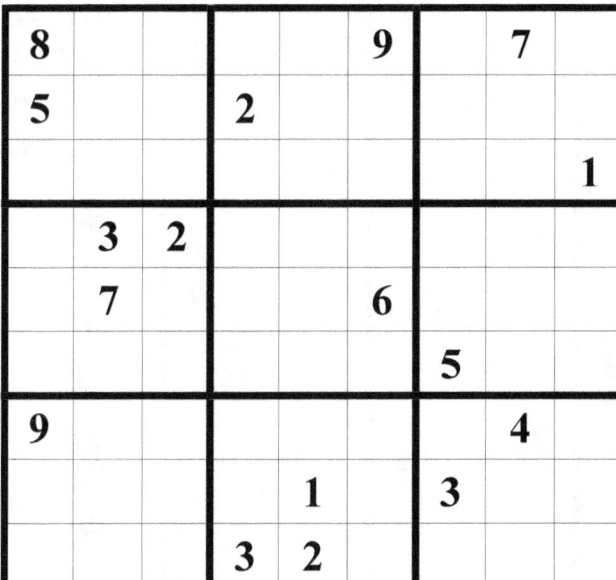

8					9		7	
5			2					
								1
	3	2						
	7				6			
						5		
9							4	
				1		3		
			3	2				

Sudoku Seventeen, by John Austin

151

★
★
★
★
★

						1		6
3	9							
						5		
			6	1		2		
8	7							
			5					
			4		7		3	
		1		5				
							8	

152

★
★
★
★
★

						2	5	
8				4				
						1		
6			2			7		
3								8
		4	1					
	2		5					
				3			6	
	1							

126

Sudoku Seventeen, by John Austin

Puzzle 1 (top):

							3		9

Top grid:

						3		9
	6							
	5							
3				8		1		
						4	6	
				2				
			7		5		2	
1			4					
			6				5	

Target: 52 min(est.)

153

Bottom grid:

					7		5	
6						8		
			3					
				9		7		2
	3	5						
	4							
8				6				
			9					4
7							3	

Target: 52 min(est.)

154

Puzzle 155 — Target: 52 min(est.)

				2			7	4
6		1						
3						1	6	
	7		5	4				
						8		
	4							8
			1		3			
			6					

155

Puzzle 156 — Target: 52 min(est.)

			2	1		5		
8							6	
	1	2						
		5					3	
	4		8					
				5		1		7
7			3					
						2		

156

Sudoku Seventeen, by John Austin

157

						3		1
7	6							
			2					
			8			6	7	
		1						
								9
	7					8	4	
5			1		3			
			5					

158

						6	9	
7							3	
			2					
2						3		8
4			6		9			
			1					
	6			3		5		
	1							
								2

Target: 52 min(est.)

159

						8		1
3	7							
	1	8	5					
4				6		3		
6						3	7	
			8	1		5		
			2					

Target: 52 min(est.)

160

					2		1	
	4					6		
	5			3				
			4	8		5		
2								3
1								
			3					8
7							2	
			5					

Sudoku Seventeen, by John Austin

161

					2	9		
3	7							
5								
8			5				3	
			1					
						7		
	2			6		1		
				8		2		4
			7					

162

				1			2	
8			5					
					7			
			2				5	8
	3	7						
						4		
				3	7	6		
5			4					
						1		

Sudoku Seventeen, by John Austin

163

★ ★ ★ ★ ★

				1	2	5		
9								
				7				
	7						8	
	2		5					
			9					4
3			6			1		
						2	7	
4								

164

★ ★ ★ ★ ★

				2			4	
7						3		
			9					
5			7		3			
	3						2	
			1			6		
	4	2		8				
								1
						7		

Sudoku Seventeen, by John Austin

Puzzle 165

Target: 52 min(est.)

165

			2			4		
	6	5						
1								
2							1	7
			1				5	
4				3				
			5		7			
3						8		
			6					

Puzzle 166

Target: 52 min(est.)

166

			2	5				3
7		1						
	2			9				
			8				7	
					6			
6	5					2		
			1	4				
8			7					

Sudoku Seventeen, by John Austin

Target: 52 min(est.)

167

★ ★ ★ ★ ★

				3	2	5		
6				1			7	
8								
	2		4		5			
			8				6	
								1
	3					2		
4			6					

Target: 52 min(est.)

168

★ ★ ★ ★ ★

				4		1		
6						5		
	3							
			2	7				6
	4						3	
		1						
2			6		3			
						4	8	
			5					

Puzzle 169

Target: 52 min(est.)

169 ★★★★★

					6	1		4
	2		7					
				5				
	7		2				8	
						7		
				4				
			8			3	5	
4		1						
6								

Puzzle 170

Target: 47 min

170 ★★★★★

			7				1	
4								
9								
				4		6		2
	1		3		9			
	7							
5						8		
				6		4		
	3		1					

Sudoku Seventeen, by John Austin

171

			8			6		
	5	4						
							2	
			4		7			
3						8		
						1		
8				6			4	
				3				5
					1		7	

172

		3		7				
							1	
							6	
			5			3		2
6					4			
	9					7		
			2	3		5		
4			8					
1								

136

Sudoku Seventeen, by John Austin

		7	5	4				
	8						2	
							1	
1		4		5		6		
			3		2			
2	3		8					
						4		
9								

173

	1							4
			8				6	
	5							
8			6			3		
						1		7
				2				
			7		1	5		
9							2	
4								

174

Target: 50 min

175

★ ★ ★ ★ ★

Target: 52 min(est.)

176

★ ★ ★ ★ ★

Sudoku Seventeen, by John Austin

Target: 47 min

177

	5					3		8
		4	7		6			
			1					
				2				3
7							4	
1								
	3			8				
			6				7	
						2		

Target: 48 min

178

	6	1						
					4	5		
				3				
	3		6	1			2	
7						4		
			9					
			2				6	
4					8			
5								

Sudoku Seventeen, by John Austin

179

★
★
★
★
★

	8		9			5		
1				6				
3			5		8			
2							9	
			4					
	4	5				8		
				2			6	
	1							

180

★
★
★
★
★

	8	4	5					
				2		7		
	5		1					
6						2		
			4		8			
7				3				
							1	8
2							9	

Sudoku Seventeen, by John Austin

181

★ ★ ★ ★ ★

4				7	3			
	6					1		
			6	1		5		
3		7					4	
			8					
					5		2	
		2						3
	1							

182

★ ★ ★ ★ ★

6	9			2				
				4			1	
								8
	5		1		7			
3						2		
			8					
7		1						
				5		9		
			3					

Sudoku Seventeen, by John Austin

183

★ ★ ★ ★ ★

Puzzle 183:

7	6					2		
			1					
2								
		8					5	9
	3			7				
				4			1	
4		1	5					
						7		6

184

★ ★ ★ ★ ★

Puzzle 184:

6		7					3	
1				2				
					7			
5			6		3			
	8							4
						2		
	2			4				
			3				6	
						1		

Sudoku Seventeen, by John Austin

1						3		
	3				4			
			2					8
				1		6	4	
	2		7					
5		8						
6				3				
								1
						2		

185

	2	7	6			5		
	3		5					
1				4	8			
				3		2		
4								
	5		2					
8								4
						1		

186

Sudoku Seventeen, by John Austin

Target: 49 min

187

★
★
★
★
★

1							4	
	8			2				
				7		2		8
5			3					
4						6		
6					4		5	
			8		3			
	7							

Target: 52 min

188

★
★
★
★
★

1			8				5	
				4			3	
			6					
	6			2		9		
				3				1
			9			4		
3						2		
5		7						

Sudoku Seventeen, by John Austin

				2			8	3
7	4		1					
	2			8	3			
						1		
		3						
5			4		6	7		
			7					
9								

189

				3	5			
		1					8	
	9							
	4		1	7				
						8	2	
			9					
			2	1				9
5						6		
8								

190

Sudoku Seventeen, by John Austin

				8		3		9
6	2							5
5				1				
			4				6	
	9						2	
			2		9	8		
1								
			7					

Target: 43 min

191

					1		2	
6	3							
			4	6				3
		8				5		
2								
	7		5	3		4		
		1					8	
			6					

Target: 43 min

192

Puzzle 193 — Target: 75 min(est.)

193

			3				4	
	2						7	
	1		5					
7							8	
			6			4		
			1					
	6				2	1		
4				7		3		

Puzzle 194 — Target: 75 min(est.)

194

				3		4	5	
1			7					
2						6		
	4					3		
8			1					
			2					
							8	1
	3		6					
	5							

Sudoku Seventeen, by John Austin

195

				3	1			
	5						4	
							9	
				2		8		
3						1		
	4		6					
7						3		
		9	4					
			5					6

196

				6		3		
1		4						
8	9							
			8		1			
	3					7		
	5							
			2					8
9							1	
			7	3				

Sudoku Seventeen, by John Austin

			2		1		7	
3	4							
5				3				
		2					1	
						8		
				6		3		5
	6	1	5					
						4		

197

			7	6		2		
4			8					
1	5							
					1		4	
	3					8		
	2							
			2					3
7							1	
			8					

198

Sudoku Seventeen, by John Austin

Puzzle 199

199

★
★★
★★
★

3			7					
			6			5		
	8							1
2						6	3	
			8	1				
5								
	7						8	4
6			2					

Puzzle 200

200

★
★★
★★
★

6			2					
5						3		
			8		4			
	7			3				
		8					4	
				6				
1						5		3
	4		7					
						6		

Sudoku Seventeen, by John Austin

201

202

Sudoku Seventeen, by John Austin

203

Target: 53 min

★
★ ★
★ ★
★

204

Target: 59 min

★
★ ★
★ ★
★

Hints for Puzzle Solutions

101 Iterate row 1, column 3 (5,9)

126 After populating the grid, row 8, column 8 must be 3 or 9. It can be shown quickly that the value 3 leads to two 3s in region 9, so row 8, column 8 is 9. Also, column 3 row 1 must (5,9) and iteration solves the puzzle.

131 Examination shows that row 6, column 5 (4,7) can't be 7 as this leads to a clash. Iterate over row 3, column 3 (6,9).

137 Iterate row 2, column 2 (7,9).

139 Iterate row 1, column 4 (1,6).

152 Iterate row 1, column 3 (7,9)

155 Iterate row 3, column 1 (1,5).

157 Iterate row 4, column 1 (2,9).

158 Iterate row 1, column 3 (5,8). This puzzle almost has two solutions!

160 Iterate row 3 column 1 (6,9).

161 Iterate row 1, column 2 (6,8).

167 Tidy up the second column.

168 Iterate row 1, column 1 (8,9).

171 Column 1 should be 2 or 9, but 2 quickly leads to a contradiction.

172 Iterate row 7, column 1 (7,9)

179 Column 5 contains 2 and 6 and the remaining 7 elements can be reduced to a linked quadruplet (1,3,7,9) and a linked triplet (4,5,8).

180 On examination you can see that row 1 column 1 (3 or 9) can only be 9.

181 Iterate for the position of the 4 in column 4 --- row 2 or 3.

182 Iterate row 2, column 1 (2,5).

187 Iterate for the position of the 5 on row 1.

188 Iterate row 6, column 6 (6,8).

189 Iterate row 1, column 3 (5,9).

Sudoku Seventeen, by John Austin

193 Iterate row 1 column 6 (8,9) and row 1 column 1 (5,8). This puzzle has almost two solutions!

194 Unravelling all the interconnections is a challenge, and the key is to spot that the 4 in region 9 is in row 8, and then 4 is identified in column 1.

195 The 6 in row 1 must be in column 2 or 3. This eliminates the 6 from other elements of region 1 and the complex interactions eventually unravel.

196 Row 9 separates into (1,8) and (2,4,5,6,9); iterate row 3, column 5 (2,7)

197 Iterate over (6,7,9) on row 2

198 Multiple iterations are needed row 3, column 4 (3,9), row 2 column 2 (6,7) and row 4, column 2 (6,9)

199 Iterations: the 4 in row 4 is in column 4 or 6. Iterate row 2, column 2 (2,9).

200 Tricky! There are a lot of dependencies to separate and then iterate row 1, column 2 (1,8).

201 Iterate for the position of the 2 in column 2 (row 8 or 9).

202 Unravelling all the interconnections is tricky. Iterate row 9, column 1 (1,7), after populating the grid.

203 There are a lot of complicated interactions and to resolve them iterate for the value of row 1, column 7 (4,5).

204 There are a lot of complex interactions and the key to this is to spot that row 7, column 2 is 3. Then iterate for row 1, column 4 (2,7).

Sudoku Seventeen, by John Austin

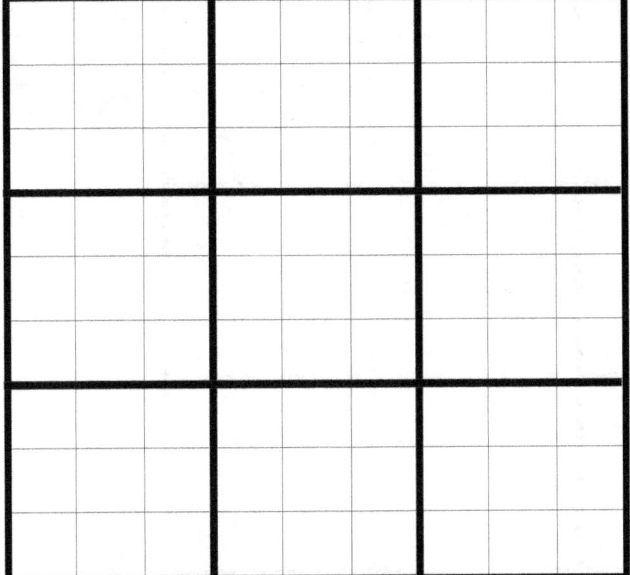

X01

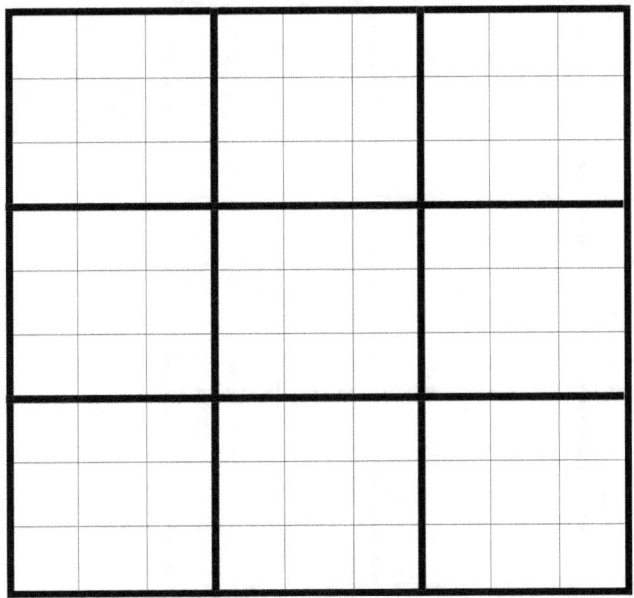

X02

Sudoku Seventeen, by John Austin

X03

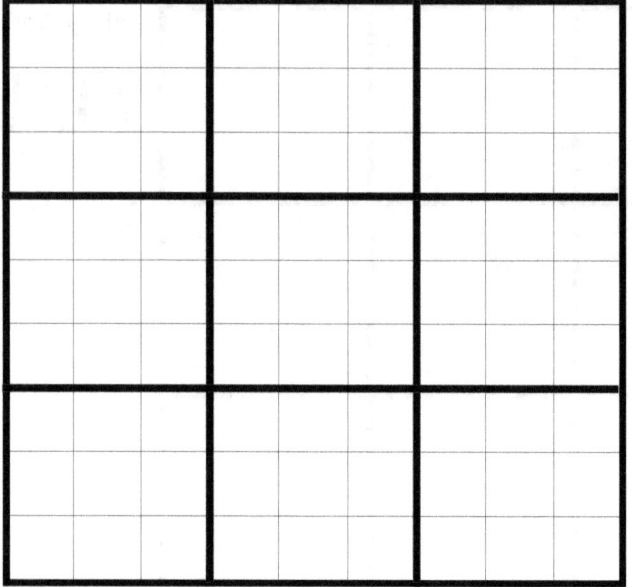

X04

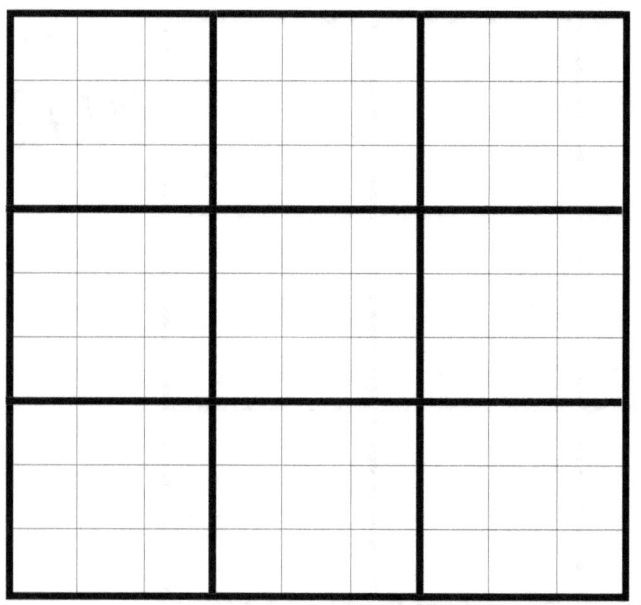

Blank Grid

X05

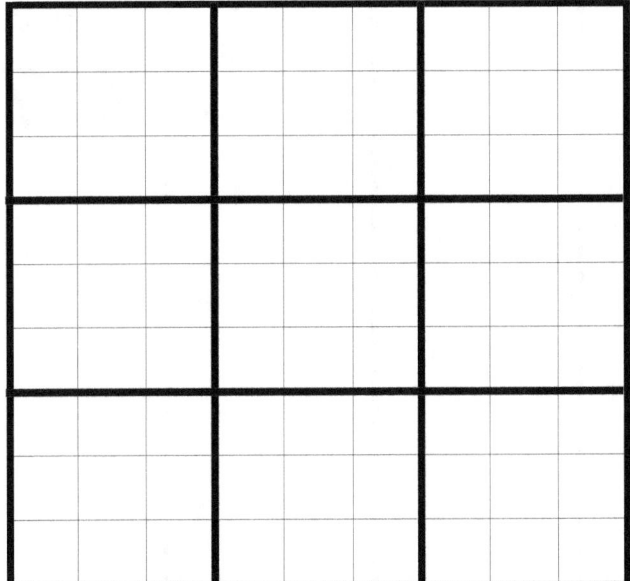

Blank grid

X06

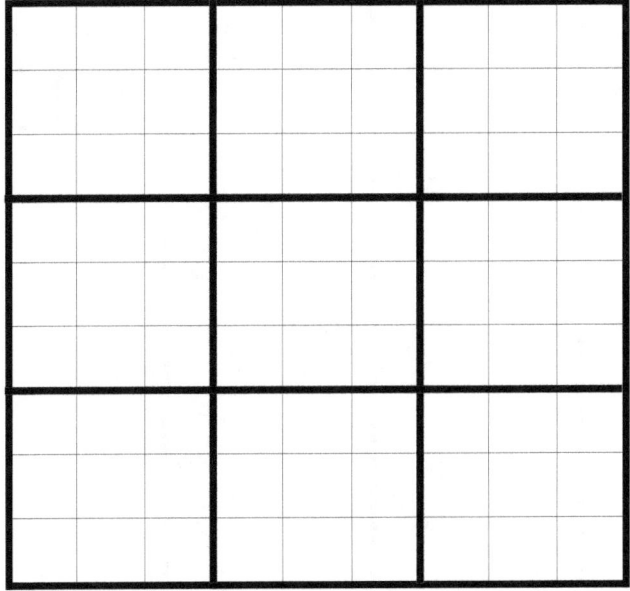

Sudoku Seventeen, by John Austin

X07

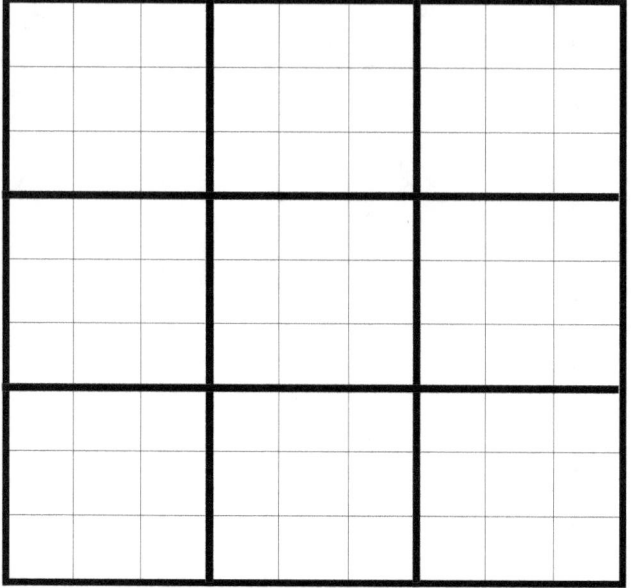

X08

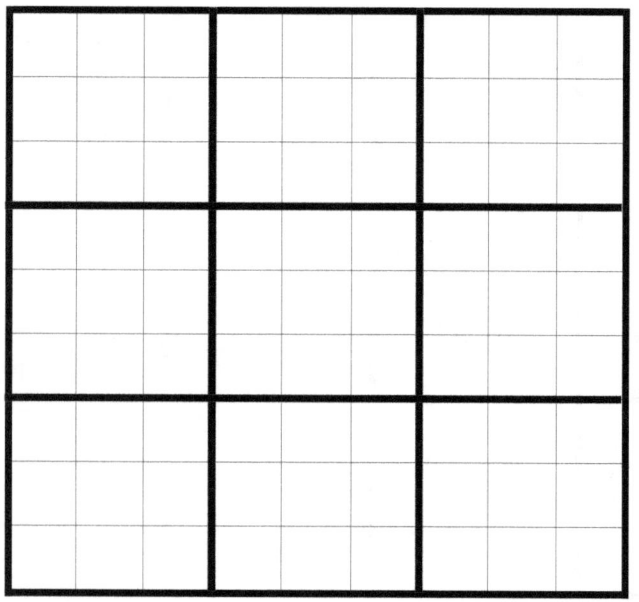

Sudoku Seventeen, by John Austin

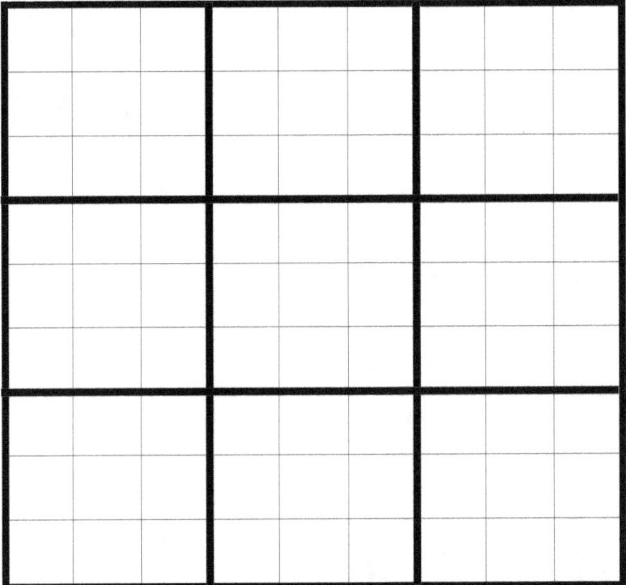

X09

X10

Sudoku Seventeen, by John Austin

X11

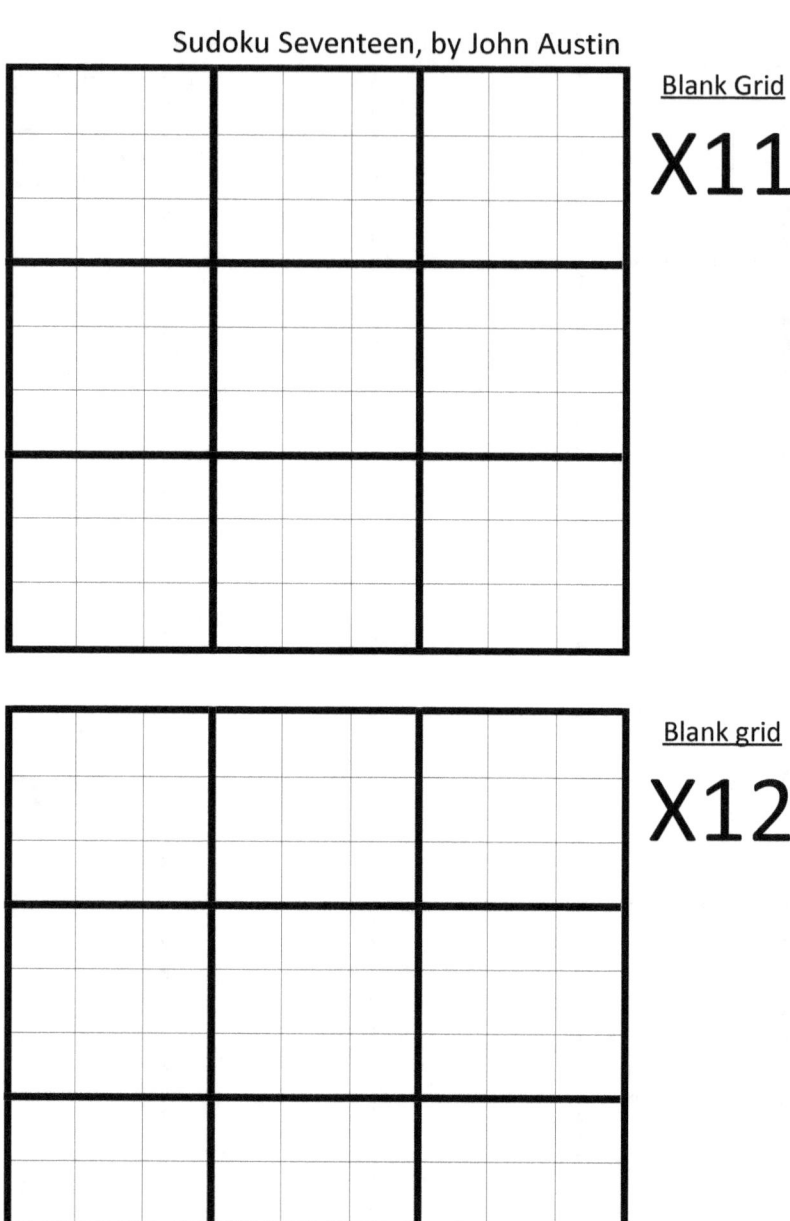

X12

Sudoku Seventeen, by John Austin

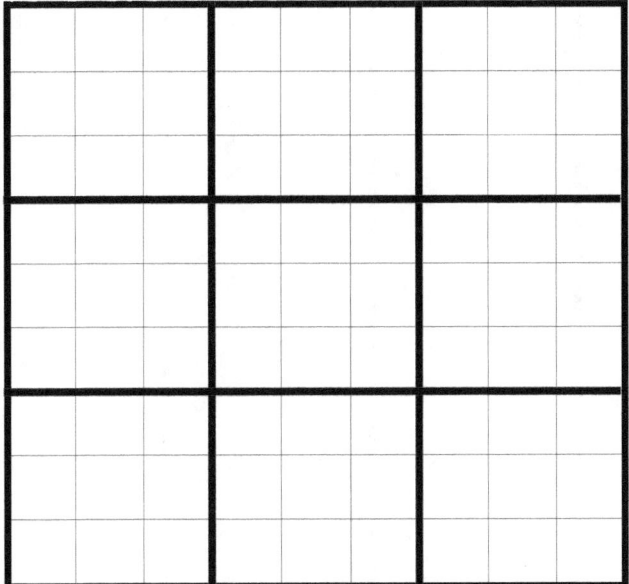

Sudoku Seventeen, by John Austin

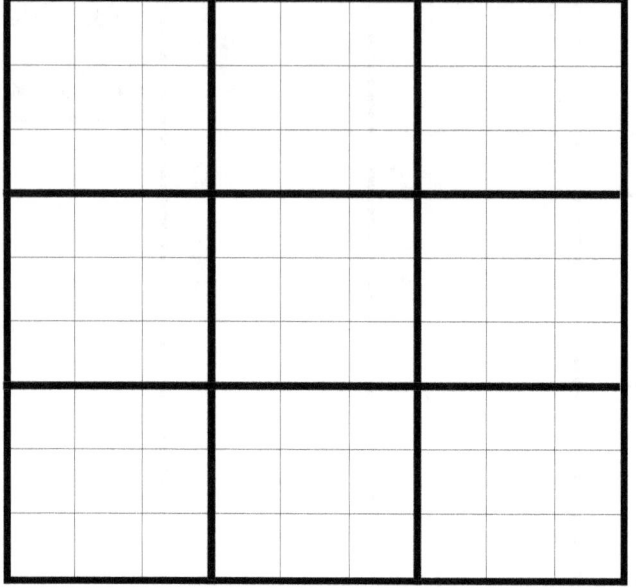

X15

X16

Sudoku Seventeen, by John Austin

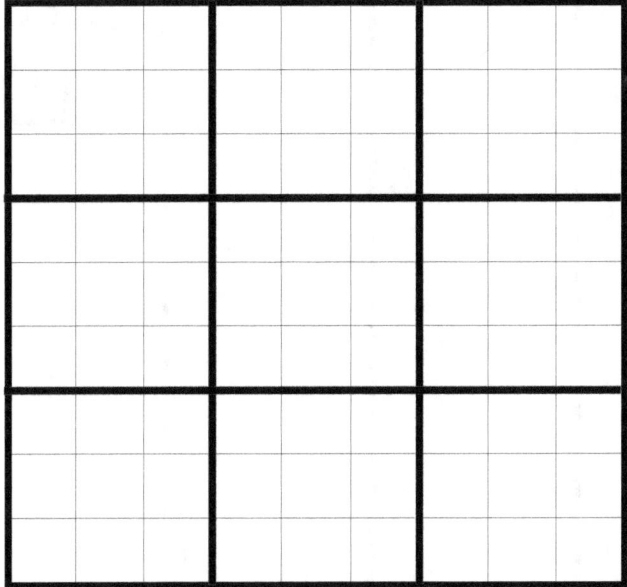

X17

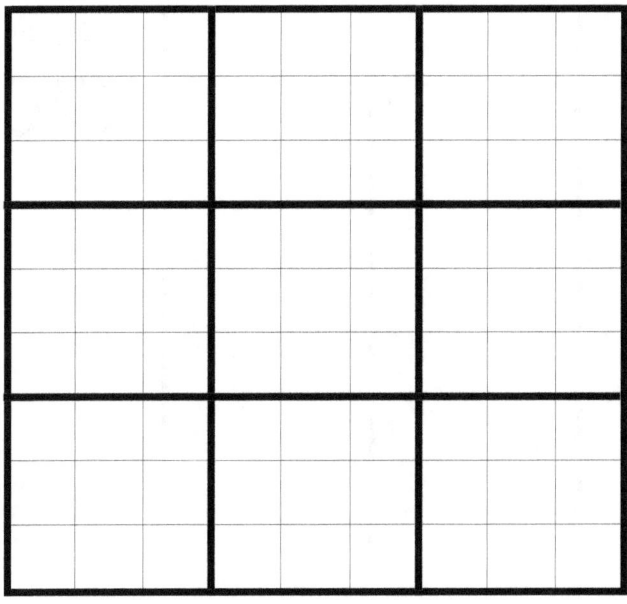

Blank grid

X18

Sudoku Seventeen, by John Austin

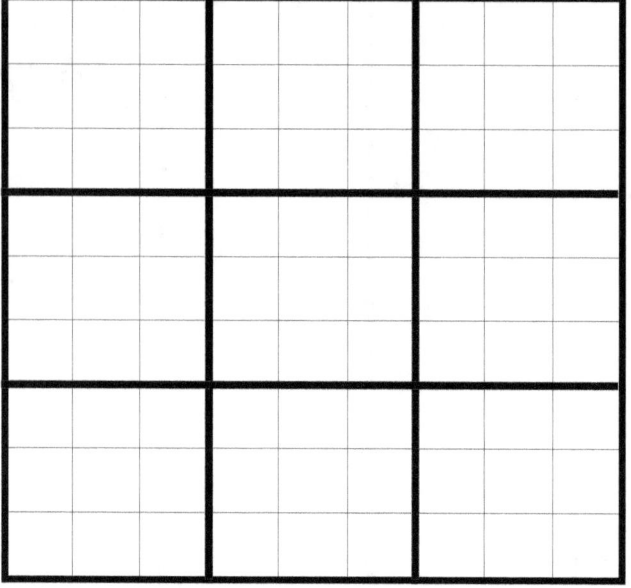

<u>Blank Grid</u>

X19

<u>Blank grid</u>

X20

The Solutions

The World's Hardest Sudoku Puzzle: Hints and Solution

Using a modified trial and error method, I managed to solve it in 2-3 days (c. 20 hours). Since I wasted some of that time, I should have done it in about 15 hours, so assuming a factor of 1.5 for each increment of difficulty class, the puzzle works out at about 12*! The primary cause of difficulty is just getting started, but with the correct early choices the solution falls out relatively quickly.

I started in region 9 where the unknown elements divide into a linked pair in column 7 and a linked triplet. The latter must contain 2 and 7 and you can guess that the third element of the triplet is 5. This would leave (3,9) in region 9, column 7. You can first explore other combinations of the triplet, and the other possible solutions fail relatively quickly. Pursuing the (2,5,7) option for the triplet, you can populate the grid. Concentrate on those elements along rows 7 and 8 which have only 2 choices. All the options except one fail and on about the 5th choice, the solution emerges, but it does take careful painstaking work and a systematic procedure for eliminating the options which fail.

8	1	2	7	5	3	6	4	9
9	4	3	6	8	2	1	7	5
6	7	5	4	9	1	2	8	3
1	5	4	2	3	7	8	9	6
3	6	9	8	4	5	7	2	1
2	8	7	1	6	9	5	3	4
5	2	1	9	7	4	3	6	8
4	3	8	5	2	6	9	1	7
7	9	6	3	1	8	4	5	2

Sudoku Seventeen, by John Austin

001 (3*)

3	5	9	2	7	6	1	4	8
7	8	1	5	9	4	3	2	6
2	4	6	8	3	1	5	7	9
1	3	5	4	8	2	9	6	7
9	6	2	3	1	7	8	5	4
4	7	8	6	5	9	2	1	3
8	2	4	1	6	3	7	9	5
6	9	3	7	2	5	4	8	1
5	1	7	9	4	8	6	3	2

002 (3*)

6	2	9	7	4	8	1	3	5
8	4	7	1	5	3	9	6	2
3	1	5	2	6	9	8	4	7
7	3	6	4	8	2	5	1	9
9	8	1	5	7	6	3	2	4
4	5	2	9	3	1	6	7	8
2	9	3	8	1	4	7	5	6
1	7	4	6	9	5	2	8	3
5	6	8	3	2	7	4	9	1

003 (3*)

6	7	4	3	5	8	9	2	1
9	5	1	7	2	4	6	3	8
2	8	3	1	6	9	5	4	7
7	2	8	5	9	1	3	6	4
1	9	6	4	8	3	2	7	5
3	4	5	2	7	6	8	1	9
8	6	2	9	4	7	1	5	3
5	1	7	8	3	2	4	9	6
4	3	9	6	1	5	7	8	2

004 (3*)

6	5	2	3	7	1	9	8	4
3	7	1	8	4	9	6	5	2
8	9	4	2	6	5	7	1	3
5	4	8	6	1	7	3	2	9
2	3	7	9	8	4	1	6	5
1	6	9	5	2	3	8	4	7
4	8	5	7	3	6	2	9	1
7	1	6	4	9	2	5	3	8
9	2	3	1	5	8	4	7	6

005 (3*)

6	2	1	9	8	3	5	7	4
9	8	4	7	2	5	3	1	6
3	5	7	1	6	4	2	8	9
5	4	6	2	1	7	9	3	8
1	7	8	4	3	9	6	2	5
2	9	3	8	5	6	1	4	7
8	3	9	5	4	1	7	6	2
4	1	5	6	7	2	8	9	3
7	6	2	3	9	8	4	5	1

006 (3*)

9	2	4	5	1	8	3	6	7
7	8	6	4	3	9	1	5	2
5	1	3	2	7	6	9	8	4
8	6	2	9	4	3	7	1	5
4	7	9	6	5	1	8	2	3
1	3	5	8	2	7	4	9	6
3	9	8	7	6	5	2	4	1
2	5	1	3	8	4	6	7	9
6	4	7	1	9	2	5	3	8

Sudoku Seventeen, by John Austin

007 (3*)

3	7	2	9	5	4	6	8	1
9	6	5	8	2	1	7	3	4
1	4	8	7	3	6	2	5	9
2	3	6	4	1	5	9	7	8
7	5	9	6	8	3	4	1	2
8	1	4	2	7	9	3	6	5
5	2	1	3	4	7	8	9	6
4	9	7	5	6	8	1	2	3
6	8	3	1	9	2	5	4	7

010 (3*)

8	3	7	9	2	1	6	4	5
2	9	1	6	4	5	7	3	8
6	5	4	7	8	3	2	1	9
3	1	8	5	9	2	4	6	7
7	2	9	4	3	6	8	5	1
5	4	6	1	7	8	3	9	2
9	7	5	8	6	4	1	2	3
1	6	3	2	5	7	9	8	4
4	8	2	3	1	9	5	7	6

008 (3*)

4	2	9	3	5	6	7	1	8
3	1	5	2	8	7	6	4	9
6	8	7	1	4	9	5	2	3
8	6	2	4	3	5	1	9	7
7	9	1	8	6	2	4	3	5
5	3	4	7	9	1	8	6	2
2	5	3	6	7	4	9	8	1
1	7	6	9	2	8	3	5	4
9	4	8	5	1	3	2	7	6

011 (3*)

3	7	5	8	2	4	1	6	9
8	1	2	9	6	5	3	7	4
6	9	4	7	1	3	2	5	8
4	5	6	1	8	2	9	3	7
1	2	7	5	3	9	4	8	6
9	3	8	4	7	6	5	2	1
5	8	1	2	9	7	6	4	3
7	4	3	6	5	1	8	9	2
2	6	9	3	4	8	7	1	5

009 (3*)

7	5	9	4	6	1	8	2	3
6	4	1	8	2	3	9	7	5
8	2	3	7	5	9	6	1	4
3	8	2	5	9	4	1	6	7
9	1	5	2	7	6	4	3	8
4	6	7	3	1	8	2	5	9
5	3	6	9	4	2	7	8	1
1	9	8	6	3	7	5	4	2
2	7	4	1	8	5	3	9	6

012 (3*)

2	9	5	8	3	6	7	4	1
7	4	3	9	5	1	2	6	8
1	6	8	7	2	4	9	3	5
4	7	1	3	6	9	5	8	2
5	3	9	2	4	8	1	7	6
6	8	2	1	7	5	4	9	3
8	1	6	5	9	7	3	2	4
3	5	7	4	8	2	6	1	9
9	2	4	6	1	3	8	5	7

Sudoku Seventeen, by John Austin

013 (3*)

8	9	5	1	4	6	2	3	7
4	1	2	5	7	3	8	9	6
6	7	3	2	8	9	4	5	1
9	5	6	4	3	8	7	1	2
3	4	1	7	9	2	5	6	8
7	2	8	6	5	1	9	4	3
1	8	4	3	2	5	6	7	9
5	6	9	8	1	7	3	2	4
2	3	7	9	6	4	1	8	5

016 (3*)

7	6	3	8	1	5	2	9	4
4	8	1	2	3	9	7	6	5
5	9	2	7	4	6	3	1	8
1	3	9	5	2	4	6	8	7
8	7	5	6	9	1	4	2	3
2	4	6	3	8	7	1	5	9
9	2	7	1	5	3	8	4	6
6	5	8	4	7	2	9	3	1
3	1	4	9	6	8	5	7	2

014 (3*)

9	4	1	3	7	8	5	6	2
6	5	7	9	1	2	8	4	3
2	3	8	6	5	4	9	7	1
4	7	2	8	9	5	1	3	6
5	1	9	2	6	3	7	8	4
8	6	3	1	4	7	2	9	5
7	9	5	4	3	1	6	2	8
3	8	6	5	2	9	4	1	7
1	2	4	7	8	6	3	5	9

017 (3*)

2	4	5	8	3	9	7	1	6
9	1	6	7	4	2	5	3	8
8	3	7	5	1	6	4	9	2
6	8	3	1	2	7	9	5	4
7	9	2	4	6	5	3	8	1
4	5	1	3	9	8	2	6	7
1	7	9	6	5	4	8	2	3
3	2	8	9	7	1	6	4	5
5	6	4	2	8	3	1	7	9

015 (3*)

5	7	3	6	2	9	8	1	4
2	9	4	8	7	1	6	3	5
1	8	6	4	5	3	2	9	7
7	1	8	9	4	5	3	2	6
3	4	2	1	6	7	9	5	8
6	5	9	3	8	2	7	4	1
9	6	7	2	1	4	5	8	3
8	3	1	5	9	6	4	7	2
4	2	5	7	3	8	1	6	9

018 (3*)

4	5	6	9	3	7	8	2	1
1	3	8	6	2	4	5	9	7
9	2	7	5	8	1	6	3	4
5	9	3	8	7	6	1	4	2
8	7	4	2	1	9	3	5	6
6	1	2	3	4	5	7	8	9
2	6	1	4	5	8	9	7	3
7	4	5	1	9	3	2	6	8
3	8	9	7	6	2	4	1	5

Sudoku Seventeen, by John Austin

019 (3*)

2	9	5	7	4	8	6	1	3
1	8	6	9	3	5	2	4	7
7	3	4	6	1	2	8	5	9
4	2	8	3	5	1	9	7	6
3	7	1	2	9	6	5	8	4
6	5	9	4	8	7	3	2	1
9	1	3	8	2	4	7	6	5
8	4	7	5	6	3	1	9	2
5	6	2	1	7	9	4	3	8

022 (3*)

8	6	7	4	5	2	9	1	3
3	5	1	6	9	8	4	2	7
2	9	4	1	3	7	8	6	5
9	7	8	3	2	1	6	5	4
5	1	3	8	6	4	2	7	9
6	4	2	9	7	5	1	3	8
7	2	9	5	8	6	3	4	1
4	8	6	7	1	3	5	9	2
1	3	5	2	4	9	7	8	6

020 (3*)

5	1	7	2	4	9	8	3	6
4	9	2	3	6	8	1	5	7
3	6	8	1	5	7	9	2	4
7	8	4	9	3	6	5	1	2
1	3	5	4	8	2	7	6	9
9	2	6	5	7	1	3	4	8
6	4	1	7	9	5	2	8	3
2	7	3	8	1	4	6	9	5
8	5	9	6	2	3	4	7	1

023 (3*)

8	7	6	2	5	3	4	9	1
2	3	1	6	4	9	7	5	8
4	9	5	1	8	7	6	2	3
6	5	7	4	9	8	3	1	2
3	2	4	5	1	6	9	8	7
1	8	9	7	3	2	5	6	4
5	1	2	9	7	4	8	3	6
9	4	8	3	6	1	2	7	5
7	6	3	8	2	5	1	4	9

021 (3*)

3	7	8	6	4	1	9	5	2
2	9	1	7	8	5	3	6	4
5	4	6	9	2	3	1	7	8
9	6	3	2	7	8	5	4	1
4	2	5	3	1	9	7	8	6
8	1	7	4	5	6	2	9	3
1	3	9	5	6	4	8	2	7
7	8	4	1	9	2	6	3	5
6	5	2	8	3	7	4	1	9

024 (3*)

2	7	6	9	5	4	3	1	8
9	1	5	7	8	3	4	6	2
3	4	8	6	2	1	7	5	9
5	6	3	4	1	9	8	2	7
7	2	1	8	3	5	9	4	6
8	9	4	2	6	7	1	3	5
1	8	2	3	9	6	5	7	4
6	5	7	1	4	8	2	9	3
4	3	9	5	7	2	6	8	1

Sudoku Seventeen, by John Austin

025 (3*)

7	6	3	2	5	1	8	4	9
1	5	8	4	3	9	6	2	7
9	4	2	8	6	7	3	1	5
4	8	6	1	7	3	5	9	2
3	2	9	5	4	8	1	7	6
5	1	7	9	2	6	4	3	8
6	7	5	3	9	4	2	8	1
2	3	1	7	8	5	9	6	4
8	9	4	6	1	2	7	5	3

028 (3*)

4	2	9	3	7	1	5	6	8
6	8	3	2	5	9	1	4	7
1	5	7	8	6	4	9	3	2
8	3	6	9	4	5	2	7	1
2	9	1	6	8	7	4	5	3
5	7	4	1	3	2	8	9	6
7	4	8	5	1	3	6	2	9
3	1	2	4	9	6	7	8	5
9	6	5	7	2	8	3	1	4

026 (3*)

2	8	4	9	5	6	1	7	3
5	9	6	7	3	1	2	8	4
1	3	7	4	8	2	9	6	5
9	7	5	6	2	4	3	1	8
3	1	2	8	7	5	6	4	9
4	6	8	1	9	3	7	5	2
8	5	1	2	6	9	4	3	7
7	4	9	3	1	8	5	2	6
6	2	3	5	4	7	8	9	1

029 (3*)

5	8	7	4	9	3	1	6	2
4	1	3	5	6	2	9	7	8
6	2	9	8	1	7	5	3	4
7	6	2	1	3	4	8	5	9
3	5	8	2	7	9	6	4	1
9	4	1	6	5	8	3	2	7
8	7	6	3	4	1	2	9	5
1	3	4	9	2	5	7	8	6
2	9	5	7	8	6	4	1	3

027 (3*)

4	7	5	8	6	3	9	1	2
9	2	1	4	5	7	8	6	3
3	8	6	1	2	9	7	5	4
6	5	7	2	9	4	3	8	1
2	9	3	6	8	1	5	4	7
8	1	4	7	3	5	6	2	9
5	6	9	3	1	2	4	7	8
1	4	8	9	7	6	2	3	5
7	3	2	5	4	8	1	9	6

030 (3*)

2	4	6	1	5	7	3	9	8
9	5	7	8	3	6	2	1	4
8	3	1	4	2	9	6	7	5
3	9	4	6	8	1	7	5	2
5	1	8	2	7	3	4	6	9
7	6	2	5	9	4	8	3	1
6	8	9	3	1	2	5	4	7
4	7	5	9	6	8	1	2	3
1	2	3	7	4	5	9	8	6

Sudoku Seventeen, by John Austin

031 (3*)

3	8	9	2	5	4	1	6	7
6	5	7	8	9	1	3	2	4
4	2	1	7	3	6	5	8	9
1	7	2	9	6	5	4	3	8
5	3	8	1	4	7	6	9	2
9	6	4	3	8	2	7	1	5
8	9	6	4	7	3	2	5	1
2	4	3	5	1	8	9	7	6
7	1	5	6	2	9	8	4	3

034 (3*)

9	3	7	4	5	1	2	6	8
6	4	5	3	8	2	9	1	7
2	1	8	6	9	7	5	3	4
1	7	4	8	2	9	3	5	6
5	2	3	7	4	6	8	9	1
8	6	9	5	1	3	7	4	2
4	9	2	1	3	8	6	7	5
3	5	6	2	7	4	1	8	9
7	8	1	9	6	5	4	2	3

032 (3*)

9	8	4	3	2	1	7	5	6
2	1	6	5	4	7	3	8	9
7	5	3	8	6	9	2	4	1
4	2	8	9	7	6	5	1	3
6	9	1	2	3	5	8	7	4
3	7	5	1	8	4	9	6	2
8	3	7	6	1	2	4	9	5
1	4	9	7	5	3	6	2	8
5	6	2	4	9	8	1	3	7

035 (3*)

6	9	7	4	1	8	5	3	2
2	8	3	5	9	7	6	1	4
5	1	4	3	2	6	8	9	7
4	3	5	8	6	9	7	2	1
9	2	1	7	5	3	4	8	6
8	7	6	1	4	2	9	5	3
1	6	8	9	3	4	2	7	5
7	5	2	6	8	1	3	4	9
3	4	9	2	7	5	1	6	8

033 (3*)

5	9	4	3	8	1	6	7	2
2	1	7	5	4	6	8	9	3
8	6	3	7	9	2	5	1	4
6	3	9	2	7	4	1	8	5
4	2	1	8	5	9	3	6	7
7	8	5	1	6	3	4	2	9
9	7	6	4	1	5	2	3	8
3	5	8	6	2	7	9	4	1
1	4	2	9	3	8	7	5	6

036 (3*)

6	5	2	7	1	3	8	4	9
4	7	1	5	9	8	3	2	6
9	8	3	2	6	4	5	1	7
5	9	4	6	3	7	2	8	1
1	3	7	9	8	2	6	5	4
8	2	6	1	4	5	9	7	3
2	4	9	8	7	6	1	3	5
3	1	5	4	2	9	7	6	8
7	6	8	3	5	1	4	9	2

Sudoku Seventeen, by John Austin

037 (3*)

5	8	1	2	4	9	6	7	3
2	7	9	3	6	5	8	1	4
4	6	3	1	8	7	9	2	5
9	3	8	5	2	4	7	6	1
7	2	5	8	1	6	4	3	9
1	4	6	9	7	3	5	8	2
6	5	7	4	3	1	2	9	8
3	9	2	6	5	8	1	4	7
8	1	4	7	9	2	3	5	6

040 (3*)

4	1	7	6	9	8	2	3	5
6	9	2	3	5	1	8	4	7
5	3	8	4	7	2	9	1	6
1	2	4	5	3	7	6	9	8
8	6	5	9	2	4	3	7	1
3	7	9	1	8	6	5	2	4
7	4	3	8	6	9	1	5	2
9	8	1	2	4	5	7	6	3
2	5	6	7	1	3	4	8	9

038 (3*)

9	1	3	5	4	7	6	8	2
2	6	8	1	3	9	5	4	7
4	7	5	8	2	6	3	8	1
7	8	4	3	9	5	1	2	6
3	2	1	6	8	4	7	5	9
6	5	9	7	1	2	4	3	8
5	4	6	9	7	8	2	1	3
8	3	7	2	5	1	9	6	4
1	9	2	4	6	3	8	7	5

041 (3*)

5	1	6	3	8	9	4	7	2
9	8	3	7	4	2	5	1	6
4	2	7	6	5	1	8	9	3
1	6	5	8	7	3	9	2	4
3	4	9	5	2	6	1	8	7
2	7	8	1	9	4	3	6	5
7	3	4	9	6	8	2	5	1
6	9	2	4	1	5	7	3	8
8	5	1	2	3	7	6	4	9

039 (3*)

4	1	6	5	7	8	9	3	2
2	3	7	1	6	9	8	5	4
5	8	9	3	4	2	1	6	7
8	7	4	2	9	5	6	1	3
9	2	3	4	1	6	5	7	8
1	6	5	7	8	3	4	2	9
6	9	2	8	5	7	3	4	1
7	5	1	9	3	4	2	8	6
3	4	8	6	2	1	7	9	5

042 (3*)

3	2	1	9	7	6	8	4	5
4	8	5	2	1	3	9	7	6
7	6	9	4	8	5	1	3	2
6	4	8	3	5	9	7	2	1
2	1	7	6	4	8	3	5	9
5	9	3	1	2	7	6	8	4
9	7	4	5	3	1	2	6	8
8	5	6	7	9	2	4	1	3
1	3	2	8	6	4	5	9	7

043 (3*)

7	2	4	8	1	6	3	5	9
5	1	9	7	4	3	2	6	8
6	3	8	5	2	9	7	4	1
1	7	3	6	8	4	9	2	5
4	8	5	9	7	2	1	3	6
2	9	6	3	5	1	8	7	4
3	6	2	4	9	8	5	1	7
9	4	7	1	3	5	6	8	2
8	5	1	2	6	7	4	9	3

046 (3*)

7	4	5	6	8	1	9	3	2
9	8	2	5	3	7	1	6	4
3	1	6	9	4	2	5	8	7
1	6	3	2	5	4	8	7	9
8	2	7	1	6	9	3	4	5
4	5	9	3	7	8	6	2	1
5	3	4	7	1	6	2	9	8
6	9	8	4	2	5	7	1	3
2	7	1	8	9	3	4	5	6

044 (3*)

5	2	8	4	9	6	1	7	3
9	3	4	2	7	1	5	8	6
7	6	1	3	8	5	4	2	9
1	7	3	9	6	8	2	5	4
2	9	6	5	4	7	8	3	1
4	8	5	1	3	2	9	6	7
6	1	7	8	2	4	3	9	5
3	4	2	6	5	9	7	1	8
8	5	9	7	1	3	6	4	2

047 (3*)

1	5	6	2	9	4	7	3	8
2	4	3	5	7	8	6	1	9
9	7	8	6	1	3	2	4	5
6	1	5	9	3	7	8	2	4
7	8	2	4	5	1	3	9	6
4	3	9	8	2	6	1	5	7
8	2	1	7	4	9	5	6	3
3	9	7	1	6	5	4	8	2
5	6	4	3	8	2	9	7	1

045 (3*)

8	4	2	5	1	9	3	6	7
9	5	7	4	3	6	8	1	2
1	6	3	8	2	7	9	5	4
4	7	1	2	9	5	6	3	8
3	8	5	7	6	1	4	2	9
6	2	9	3	4	8	5	7	1
2	1	4	6	8	3	7	9	5
5	9	6	1	7	4	2	8	3
7	3	8	9	5	2	1	4	6

048 (3*)

9	6	1	3	4	5	8	2	7
2	8	5	6	7	1	9	3	4
4	3	7	2	8	9	1	5	6
5	1	8	7	9	4	3	6	2
3	2	4	8	1	6	7	9	5
6	7	9	5	3	2	4	8	1
8	4	6	9	5	7	2	1	3
1	5	3	4	2	8	6	7	9
7	9	2	1	6	3	5	4	8

Sudoku Seventeen, by John Austin

049 (3*)

7	6	1	2	8	4	3	9	5
2	9	8	3	7	5	4	1	6
4	3	5	6	9	1	8	7	2
5	8	4	9	1	6	7	2	3
9	2	3	8	4	7	6	5	1
6	1	7	5	3	2	9	4	8
1	5	9	7	6	8	2	3	4
3	4	6	1	2	9	5	8	7
8	7	2	4	5	3	1	6	9

052 (3*)

6	8	4	5	1	9	7	2	3
3	2	5	4	7	6	9	1	8
9	7	1	8	3	2	4	6	5
8	3	6	2	5	7	1	4	9
7	1	2	9	4	3	5	8	6
4	5	9	6	8	1	2	3	7
2	4	3	7	9	8	6	5	1
5	9	8	1	6	4	3	7	2
1	6	7	3	2	5	8	9	4

050 (3*)

9	7	1	2	4	6	8	5	3
6	5	2	3	8	1	9	7	4
8	4	3	5	9	7	1	2	6
4	1	9	6	5	2	3	8	7
3	2	6	9	7	8	5	4	1
7	8	5	1	3	4	2	6	9
1	3	8	4	6	5	7	9	2
5	9	4	7	2	3	6	1	8
2	6	7	8	1	9	4	3	5

053 (3*)

8	9	6	1	4	7	2	5	3
1	7	5	3	2	6	9	8	4
4	2	3	9	5	8	7	6	1
3	1	7	6	9	2	5	4	8
9	6	2	4	8	5	3	1	7
5	8	4	7	1	3	6	9	2
6	5	8	2	3	1	4	7	9
7	3	9	8	6	4	1	2	5
2	4	1	5	7	9	8	3	6

051 (3*)

9	7	1	5	4	6	8	2	3
6	5	4	2	8	3	7	9	1
8	2	3	1	9	7	6	5	4
2	1	8	7	5	4	9	3	6
7	4	6	8	3	9	5	1	2
5	3	9	6	1	2	4	8	7
4	6	2	3	7	8	1	9	5
3	9	5	4	6	1	2	7	8
1	8	7	9	2	5	3	6	4

054 (3*)

1	5	3	6	7	4	9	2	8
4	6	2	8	9	1	5	7	3
7	9	8	3	2	5	4	1	6
5	8	7	9	1	6	3	4	2
6	3	9	2	4	8	7	5	1
2	1	4	7	5	3	8	6	9
3	4	1	5	8	2	6	9	7
8	7	5	1	6	9	2	3	4
9	2	6	4	3	7	1	8	5

Sudoku Seventeen, by John Austin

055 (3*)

1	8	7	6	4	9	3	2	5
5	3	6	2	1	8	4	9	7
2	9	4	7	3	5	6	1	8
6	5	9	1	7	3	2	8	4
4	2	8	9	5	6	7	3	1
7	1	3	8	2	4	9	5	6
8	4	2	3	6	1	5	7	9
9	7	5	4	8	2	1	6	3
3	6	1	5	9	7	8	4	2

058 (3*)

2	4	1	3	9	7	6	5	8
8	6	9	2	5	1	4	3	7
7	3	5	8	4	6	1	9	2
1	8	6	7	3	4	9	2	5
3	5	4	9	1	2	7	8	6
9	2	7	5	6	8	3	1	4
6	7	3	1	8	5	2	4	9
4	1	8	6	2	9	5	7	3
5	9	2	4	7	3	8	6	1

056 (3*)

1	6	5	2	8	7	9	4	3
8	7	2	9	3	4	1	5	6
4	9	3	6	1	5	7	2	8
2	3	8	5	6	1	4	7	9
7	4	9	3	2	8	6	1	5
5	1	6	7	4	9	8	3	2
9	2	1	4	5	6	3	8	7
3	8	7	1	9	2	5	6	4
6	5	4	8	7	3	2	9	1

059 (3*)

2	9	7	4	3	6	5	8	1
1	6	8	7	9	5	3	4	2
4	5	3	1	2	8	7	9	6
7	3	4	9	5	1	2	6	8
9	8	5	2	6	4	1	7	3
6	2	1	3	8	7	4	5	9
3	1	6	5	4	9	8	2	7
8	4	2	6	7	3	9	1	5
5	7	9	8	1	2	6	3	4

057 (3*)

2	7	4	8	9	1	5	3	6
6	3	8	7	4	5	9	2	1
1	9	5	6	3	2	7	8	4
7	5	2	4	8	9	1	6	3
3	4	9	2	1	6	8	7	5
8	6	1	3	5	7	4	9	2
5	8	7	1	2	3	6	4	9
4	1	3	9	6	8	2	5	7
9	2	6	5	7	4	3	1	8

060 (3*)

3	9	6	7	2	5	4	8	1
2	7	4	1	8	6	9	3	5
1	5	8	9	3	4	6	7	2
9	8	3	2	6	1	7	5	4
6	4	2	5	7	8	3	1	9
5	1	7	3	4	9	8	2	6
7	6	9	8	1	2	5	4	3
4	3	1	6	5	7	2	9	8
8	2	5	4	9	3	1	6	7

Sudoku Seventeen, by John Austin

061 (3*)

4	3	7	2	1	8	6	5	9
9	5	2	7	3	6	8	4	1
8	6	1	5	4	9	3	2	7
2	8	5	9	7	4	1	3	6
1	9	3	6	2	5	7	8	4
7	4	6	1	8	3	5	9	2
6	2	9	8	5	1	4	7	3
5	1	4	3	9	7	2	6	8
3	7	8	4	6	2	9	1	5

064 (3*)

5	9	2	6	8	1	4	3	7
3	6	1	2	4	7	9	5	8
8	7	4	5	3	9	1	6	2
4	1	3	8	7	5	2	9	6
9	5	6	4	1	2	8	7	3
7	2	8	9	6	3	5	1	4
1	3	9	7	2	4	6	8	5
6	4	5	3	9	8	7	2	1
2	8	7	1	5	6	3	4	9

062 (3*)

4	6	3	2	9	1	5	8	7
2	5	9	8	7	3	1	6	4
1	7	8	6	4	5	3	9	2
7	1	4	9	5	6	8	2	3
5	3	2	4	1	8	9	7	6
8	9	6	7	3	2	4	1	5
3	2	7	5	8	9	6	4	1
9	4	1	3	6	7	2	5	8
6	8	5	1	2	4	7	3	9

065 (3*)

5	2	3	9	6	1	8	7	4
4	6	7	3	2	8	1	5	9
1	8	9	4	7	5	2	3	6
3	9	6	2	1	4	5	8	7
2	4	8	7	5	9	3	6	1
7	1	5	6	8	3	4	9	2
8	7	1	5	9	2	6	4	3
9	5	4	1	3	6	7	2	8
6	3	2	8	4	7	9	1	5

063 (3*)

5	4	8	9	2	1	7	3	6
9	1	2	6	7	3	4	5	8
6	3	7	5	8	4	2	9	1
2	9	5	7	3	8	6	1	4
3	6	1	4	5	2	8	7	9
8	7	4	1	9	6	5	2	3
1	8	9	2	6	5	3	4	7
4	5	6	3	1	7	9	8	2
7	2	3	8	4	9	1	6	5

066 (3*)

5	3	6	2	9	8	4	7	1
1	9	7	5	6	4	2	3	8
2	8	4	7	3	1	5	6	9
9	4	5	3	2	6	1	8	7
6	7	3	1	8	5	9	4	2
8	2	1	4	7	9	3	5	6
4	1	8	9	5	7	6	2	3
7	5	2	6	1	3	8	9	4
3	6	9	8	4	2	7	1	5

Sudoku Seventeen, by John Austin

067 (3*)

6	8	3	5	2	9	4	7	1
7	4	5	3	1	8	9	6	2
9	2	1	6	4	7	3	5	8
4	5	8	7	6	2	1	9	3
3	1	6	4	9	5	8	2	7
2	9	7	1	8	3	5	4	6
1	7	4	8	5	6	2	3	9
8	6	9	2	3	4	7	1	5
5	3	2	9	7	1	6	8	4

070 (3*)

7	6	3	1	8	4	5	9	2
1	5	9	7	3	2	6	8	4
4	2	8	9	6	5	1	3	7
3	4	2	8	1	7	9	5	6
5	8	7	6	4	9	2	1	3
9	1	6	2	5	3	7	4	8
2	9	1	4	7	8	3	6	5
8	7	5	3	9	6	4	2	1
6	3	4	5	2	1	8	7	9

068 (3*)

6	9	4	7	5	3	2	1	8
7	2	1	8	6	9	4	5	3
8	3	5	4	2	1	9	6	7
1	4	3	5	9	8	7	2	6
5	7	6	2	1	4	8	3	9
9	8	2	6	3	7	1	4	5
3	5	9	1	7	2	6	8	4
4	1	7	3	8	6	5	9	2
2	6	8	9	4	5	3	7	1

071 (3*)

7	8	3	6	4	5	1	9	2
1	9	5	2	3	7	6	8	4
4	2	6	8	9	1	7	3	5
6	7	9	3	2	4	5	1	8
3	1	8	7	5	9	4	2	6
5	4	2	1	8	6	9	7	3
2	6	1	5	7	8	3	4	9
9	3	7	4	6	2	8	5	1
8	5	4	9	1	3	2	6	7

069 (3*)

7	8	9	4	2	1	6	5	3
4	5	6	8	9	3	2	1	7
1	2	3	6	7	5	8	9	4
8	6	1	2	4	9	7	3	5
3	7	5	1	6	8	4	2	9
9	4	2	5	3	7	1	8	6
5	9	4	7	8	2	3	6	1
6	1	8	3	5	4	9	7	2
2	3	7	9	1	6	5	4	8

072 (3*)

8	2	9	5	3	4	6	7	1
7	1	4	9	2	6	5	8	3
5	3	6	7	1	8	9	2	4
9	4	8	6	5	1	2	3	7
3	5	1	4	7	2	8	6	9
2	6	7	3	8	9	1	4	5
4	8	5	2	9	3	7	1	6
1	9	3	8	6	7	4	5	2
6	7	2	1	4	5	3	9	8

Sudoku Seventeen, by John Austin

073 (3*)

8	6	1	5	7	3	2	4	9
9	7	2	6	4	8	5	1	3
4	5	3	1	2	9	8	7	6
3	9	4	8	1	7	6	5	2
2	1	5	3	6	4	9	8	7
6	8	7	9	5	2	4	3	1
5	3	6	4	9	1	7	2	8
1	2	9	7	8	5	3	6	4
7	4	8	2	3	6	1	9	5

076 (3*)

7	6	3	5	8	1	9	4	2
8	2	5	4	3	9	7	1	6
9	1	4	6	2	7	8	5	3
5	9	1	2	6	8	3	7	4
6	3	7	1	9	4	2	8	5
4	8	2	7	5	3	6	9	1
1	7	8	3	4	2	5	6	9
2	4	6	9	7	5	1	3	8
3	5	9	8	1	6	4	2	7

074 (3*)

8	3	4	9	7	6	5	2	1
9	7	5	1	8	2	3	4	6
6	1	2	4	5	3	8	9	7
7	5	8	2	1	9	6	3	4
4	2	3	8	6	5	7	1	9
1	9	6	3	4	7	2	5	8
2	6	1	5	9	8	4	7	3
5	4	7	6	3	1	9	8	2
3	8	9	7	2	4	1	6	5

077 (3*)

4	1	7	3	9	8	2	6	5
6	2	8	1	5	4	9	3	7
3	9	5	7	6	2	8	4	1
9	5	4	2	3	7	6	1	8
2	3	1	9	8	6	7	5	4
7	8	6	4	1	5	3	9	2
8	4	3	6	7	1	5	2	9
1	7	9	5	2	3	4	8	6
5	6	2	8	4	9	1	7	3

075 (3*)

9	7	3	5	6	1	4	2	8
1	8	5	2	9	4	6	7	3
2	6	4	7	8	3	5	9	1
8	4	2	3	7	5	9	1	6
5	3	9	1	2	6	7	8	4
7	1	6	9	4	8	2	3	5
3	2	8	6	5	9	1	4	7
4	5	7	8	1	2	3	6	9
6	9	1	4	3	7	8	5	2

078 (3*)

5	3	8	7	6	1	2	4	9
6	2	9	8	5	4	3	7	1
4	1	7	9	3	2	8	6	5
9	7	2	3	1	6	4	5	8
1	8	4	5	9	7	6	2	3
3	6	5	2	4	8	1	9	7
7	4	3	1	2	5	9	8	6
2	5	1	6	8	9	7	3	4
8	9	6	4	7	3	5	1	2

Sudoku Seventeen, by John Austin

079 (4*)

9	8	6	4	3	7	2	1	5
1	7	4	8	5	2	6	9	3
5	3	2	9	1	6	8	7	4
6	4	3	5	7	9	1	2	8
8	2	5	1	6	4	7	3	9
7	9	1	2	8	3	5	4	6
4	1	8	7	9	5	3	6	2
3	5	9	6	2	1	4	8	7
2	6	7	3	4	8	9	5	1

082 (4*)

5	6	9	3	2	1	7	4	8
7	8	2	6	9	4	1	5	3
4	3	1	7	8	5	6	2	9
6	4	5	9	3	7	2	8	1
3	2	7	1	4	8	5	9	6
1	9	8	5	6	2	4	3	7
2	5	6	8	1	9	3	7	4
8	7	3	4	5	6	9	1	2
9	1	4	2	7	3	8	6	5

080 (4*)

6	3	2	8	5	7	4	9	1
7	5	4	2	1	9	3	8	6
9	8	1	4	3	6	7	2	5
5	2	8	7	6	1	9	3	4
1	4	6	9	2	3	8	5	7
3	7	9	5	4	8	6	1	2
2	1	7	3	8	4	5	6	9
4	6	3	1	9	5	2	7	8
8	9	5	6	7	2	1	4	3

083 (4*)

5	8	9	4	2	1	7	6	3
1	3	2	7	6	9	8	4	5
6	4	7	5	3	8	2	1	9
7	9	6	1	8	5	4	3	2
8	2	1	3	4	6	5	9	7
3	5	4	9	7	2	1	8	6
9	7	8	6	5	4	3	2	1
2	1	5	8	9	3	6	7	4
4	6	3	2	1	7	9	5	8

081 (4*)

3	7	5	2	1	9	6	4	8
9	4	6	3	7	8	2	5	1
2	8	1	6	4	5	7	3	9
5	3	8	4	9	6	1	7	2
1	9	7	8	5	2	3	6	4
6	2	4	7	3	1	8	9	5
7	6	9	1	8	4	5	2	3
4	1	3	5	2	7	9	8	6
8	5	2	9	6	3	4	1	7

084 (4*)

7	4	6	9	5	1	2	3	8
2	8	3	6	4	7	9	5	1
5	9	1	8	2	3	4	6	7
6	2	8	1	3	4	7	9	5
3	5	9	7	6	2	8	1	4
4	1	7	5	9	8	3	2	6
1	3	4	2	8	5	6	7	9
8	6	5	3	7	9	1	4	2
9	7	2	4	1	6	5	8	3

Sudoku Seventeen, by John Austin

085 (4*)

2	8	1	6	7	3	4	9	5
9	5	6	1	2	4	7	3	8
4	3	7	9	8	5	2	1	6
8	1	2	5	9	7	3	6	4
3	6	9	4	1	8	5	2	7
5	7	4	2	3	6	1	8	9
1	4	5	3	6	9	8	7	2
6	2	8	7	4	1	9	5	3
7	9	3	8	5	2	6	4	1

088 (4*)

9	5	8	7	1	2	4	3	6
3	6	4	8	5	9	2	7	1
1	2	7	6	3	4	9	8	5
8	4	1	2	6	3	7	5	9
7	9	5	1	4	8	6	2	3
6	3	2	9	7	5	8	1	4
2	1	6	3	9	7	5	4	8
5	8	3	4	2	6	1	9	7
4	7	9	5	8	1	3	6	2

086 (4*)

6	3	4	1	7	8	5	2	9
8	9	2	4	6	5	3	1	7
1	5	7	9	3	2	4	6	8
7	1	3	5	8	6	2	9	4
2	4	8	3	1	9	6	7	5
5	6	9	7	2	4	1	8	3
4	8	5	2	9	1	7	3	6
3	2	6	8	5	7	9	4	1
9	7	1	6	4	3	8	5	2

089 (4*)

4	5	8	7	1	6	9	2	3
2	1	3	9	4	8	7	5	6
9	6	7	5	3	2	1	8	4
7	4	2	8	6	9	5	3	1
6	3	5	1	2	7	4	9	8
8	9	1	3	5	4	6	7	2
3	8	4	6	7	5	2	1	9
1	7	6	2	9	3	8	4	5
5	2	9	4	8	1	3	6	7

087 (4*)

7	1	2	9	8	5	3	6	4
9	3	6	1	2	4	8	7	5
4	5	8	7	6	3	9	2	1
8	2	1	3	7	6	5	4	9
5	6	9	4	1	2	7	8	3
3	4	7	5	9	8	2	1	6
2	7	5	6	3	1	4	9	8
6	9	3	8	4	7	1	5	2
1	8	4	2	5	9	6	3	7

090 (4*)

4	6	2	9	1	8	5	7	3
7	3	8	5	2	6	4	9	1
1	9	5	4	3	7	8	2	6
8	5	3	6	7	4	9	1	2
9	7	4	1	5	2	6	3	8
2	1	6	3	8	9	7	5	4
3	4	9	7	6	1	2	8	5
6	2	1	8	9	5	3	4	7
5	8	7	2	4	3	1	6	9

Sudoku Seventeen, by John Austin

091 (4*)

5	7	8	3	2	9	4	1	6
6	4	9	5	7	1	2	8	3
1	3	2	4	6	8	7	5	9
7	2	6	9	8	3	5	4	1
4	8	3	7	1	5	6	9	2
9	1	5	6	4	2	8	3	7
2	5	4	1	9	7	3	6	8
3	9	7	8	5	6	1	2	4
8	6	1	2	3	4	9	7	5

094 (4*)

5	2	6	8	4	1	7	3	9
7	8	1	2	9	3	4	6	5
3	9	4	7	6	5	2	1	8
6	4	3	5	7	8	1	9	2
9	1	5	4	3	2	6	8	7
8	7	2	9	1	6	3	5	4
2	6	9	3	8	4	5	7	1
1	5	7	6	2	9	8	4	3
4	3	8	1	5	7	9	2	6

092 (4*)

6	1	7	4	3	5	2	8	9
2	4	5	6	9	8	7	1	3
3	9	8	2	7	1	5	4	6
5	8	3	7	6	4	9	2	1
1	2	9	5	8	3	6	7	4
4	7	6	9	1	2	3	5	8
8	6	4	3	5	7	1	9	2
9	5	2	1	4	6	8	3	7
7	3	1	8	2	9	4	6	5

095 (4*)

9	1	5	6	4	2	3	8	7
2	3	7	8	1	9	5	4	6
8	6	4	5	3	7	9	1	2
6	8	3	9	5	1	2	7	4
5	7	1	2	6	4	8	3	9
4	9	2	7	8	3	1	6	5
1	2	6	3	7	5	4	9	8
3	5	8	4	9	6	7	2	1
7	4	9	1	2	8	6	5	3

093 (4*)

6	9	5	3	4	8	7	2	1
8	3	1	2	7	6	5	4	9
7	2	4	5	1	9	6	3	8
3	4	9	8	5	1	2	6	7
2	1	7	4	6	3	9	8	5
5	8	6	7	9	2	4	1	3
1	6	2	9	3	5	8	7	4
4	5	3	6	8	7	1	9	2
9	7	8	1	2	4	3	5	6

096 (4*)

2	1	7	9	4	6	5	8	3
8	9	4	5	3	7	1	2	6
6	3	5	8	2	1	4	9	7
9	8	6	3	1	5	7	4	2
5	2	3	4	7	9	6	1	8
7	4	1	2	6	8	3	5	9
4	6	9	7	5	2	8	3	1
3	7	2	1	8	4	9	6	5
1	5	8	6	9	3	2	7	4

Sudoku Seventeen, by John Austin

097 (4*)

9	8	3	7	5	1	2	6	4
4	2	5	6	3	8	1	7	9
1	6	7	9	2	4	8	3	5
8	1	6	4	9	5	7	2	3
5	7	2	3	1	6	4	9	8
3	4	9	8	7	2	6	5	1
7	9	1	2	4	3	5	8	6
6	3	4	5	8	7	9	1	2
2	5	8	1	6	9	3	4	7

100 (4*)

9	1	7	6	8	5	3	2	4
5	2	6	1	4	3	9	7	8
4	3	8	9	2	7	6	5	1
8	6	5	7	3	1	4	9	2
7	9	1	4	6	2	8	3	5
2	4	3	5	9	8	1	6	7
1	8	2	3	5	9	7	4	6
3	7	4	2	1	6	5	8	9
6	5	9	8	7	4	2	1	3

098 (4*)

7	9	8	5	6	2	4	3	1
4	2	1	7	3	8	5	6	9
6	5	3	1	4	9	8	7	2
2	1	7	6	5	3	9	4	8
3	4	5	8	9	7	2	1	6
8	6	9	2	1	4	3	5	7
9	3	6	4	8	1	7	2	5
5	7	4	9	2	6	1	8	3
1	8	2	3	7	5	6	9	4

101 (4*)

3	7	5	1	9	6	2	8	4
1	4	9	8	2	7	6	5	3
8	2	6	4	5	3	7	1	9
4	1	7	5	6	8	9	3	2
2	5	3	9	7	4	8	6	1
6	9	8	3	1	2	4	7	5
5	3	4	6	8	9	1	2	7
7	6	1	2	4	5	3	9	8
9	8	2	7	3	1	5	4	6

099 (4*)

1	8	4	9	7	6	5	2	3
2	9	3	8	5	1	7	6	4
6	5	7	2	4	3	9	1	8
9	7	2	3	8	4	6	5	1
3	1	8	5	6	9	2	4	7
4	6	5	7	1	2	3	8	9
7	2	1	6	3	8	4	9	5
5	4	9	1	2	7	8	3	6
8	3	6	4	9	5	1	7	2

102 (4*)

1	6	7	3	5	9	4	8	2
4	5	2	7	1	8	9	6	3
3	8	9	4	2	6	1	5	7
2	7	4	9	6	3	8	1	5
6	9	3	5	8	1	7	2	4
8	1	5	2	4	7	3	9	6
5	2	8	1	7	4	6	3	9
7	3	6	8	9	5	2	4	1
9	4	1	6	3	2	5	7	8

103 (4*)

9	6	3	5	2	7	4	8	1
2	4	1	8	3	9	5	6	7
5	8	7	1	4	6	3	2	9
4	5	6	3	1	2	9	7	8
7	9	8	6	5	4	1	3	2
3	1	2	7	9	8	6	5	4
8	7	5	4	6	1	2	9	3
1	3	9	2	8	5	7	4	6
6	2	4	9	7	3	8	1	5

106 (4*)

3	7	8	6	9	5	4	2	1
5	9	1	7	4	2	6	3	8
4	2	6	3	8	1	9	5	7
9	5	3	4	1	7	2	8	6
6	8	2	9	5	3	1	7	4
1	4	7	8	2	6	5	9	3
2	3	9	1	6	8	7	4	5
8	6	5	2	7	4	3	1	9
7	1	4	5	3	9	8	6	2

104 (4*)

9	8	4	5	1	2	6	7	3
3	7	1	8	6	9	4	5	2
5	6	2	7	3	4	1	9	8
2	9	6	1	7	8	5	3	4
8	3	7	4	2	5	9	1	6
1	4	5	6	9	3	8	2	7
7	5	8	3	4	1	2	6	9
4	2	3	9	5	6	7	8	1
6	1	9	2	8	7	3	4	5

107 (4*)

5	9	2	6	4	8	1	7	3
4	3	7	9	1	5	2	8	6
6	8	1	7	3	2	4	9	5
3	7	8	1	2	6	9	5	4
2	5	4	3	8	9	7	6	1
1	6	9	4	5	7	3	2	8
8	4	6	2	9	3	5	1	7
9	1	5	8	7	4	6	3	2
7	2	3	5	6	1	8	4	9

105 (4*)

5	1	2	6	9	3	8	7	4
7	3	6	2	4	8	5	1	9
9	8	4	5	7	1	6	2	3
8	5	9	1	6	7	4	3	2
2	6	7	3	5	4	9	8	1
1	4	3	8	2	9	7	6	5
3	7	1	9	8	5	2	4	6
6	9	8	4	1	2	3	5	7
4	2	5	7	3	6	1	9	8

108 (4*)

5	7	6	9	1	4	8	3	2
9	4	1	8	2	3	7	5	6
2	8	3	6	7	5	9	4	1
4	3	7	1	6	2	5	8	9
6	5	2	3	9	8	1	7	4
8	1	9	5	4	7	2	6	3
1	2	5	4	8	6	3	9	7
3	9	4	7	5	1	6	2	8
7	6	8	2	3	9	4	1	5

Sudoku Seventeen, by John Austin

109 (4*)

3	7	2	5	6	8	4	1	9
1	8	9	2	4	3	6	5	7
5	6	4	7	1	9	3	2	8
2	9	1	6	3	5	8	7	4
7	4	6	1	8	2	5	9	3
8	5	3	4	9	7	1	6	2
6	1	8	9	2	4	7	3	5
9	3	5	8	7	1	2	4	6
4	2	7	3	5	6	9	8	1

112 (4*)

7	2	1	9	8	4	5	6	3
8	5	3	1	6	7	4	2	9
9	4	6	3	5	2	1	8	7
1	8	2	5	3	9	7	4	6
4	3	9	7	1	6	8	5	2
6	7	5	2	4	8	3	9	1
5	6	7	8	2	1	9	3	4
3	1	4	6	9	5	2	7	8
2	9	8	4	7	3	6	1	5

110 (4*)

1	3	5	9	8	6	2	7	4
8	2	4	7	3	5	9	6	1
6	7	9	2	4	1	5	8	3
7	1	3	6	5	4	8	2	9
4	9	6	8	2	7	3	1	5
2	5	8	1	9	3	7	4	6
3	6	2	4	7	9	1	5	8
9	4	7	5	1	8	6	3	2
5	8	1	3	6	2	4	9	7

113 (4*)

9	3	1	2	6	7	5	8	4
4	5	7	8	1	3	2	9	6
8	2	6	4	9	5	7	1	3
5	8	2	1	3	6	9	4	7
1	7	3	9	4	2	6	5	8
6	4	9	5	7	8	3	2	1
3	9	8	7	2	4	1	6	5
2	6	4	3	5	1	8	7	9
7	1	5	6	8	9	4	3	2

111 (4*)

9	5	6	1	4	8	3	7	2
8	2	3	7	5	6	9	4	1
1	4	7	2	9	3	8	6	5
3	8	5	6	1	4	7	2	9
7	6	9	5	8	2	4	1	3
2	1	4	3	7	9	5	8	6
6	9	8	4	2	5	1	3	7
5	7	2	8	3	1	6	9	4
4	3	1	9	6	7	2	5	8

114 (4*)

9	3	1	8	6	2	5	4	7
8	2	4	3	7	5	9	6	1
5	6	7	1	9	4	2	8	3
1	4	5	9	2	3	8	7	6
3	9	6	7	4	8	1	5	2
2	7	8	6	5	1	3	9	4
6	1	9	5	3	7	4	2	8
4	5	3	2	8	6	7	1	9
7	8	2	4	1	9	6	3	5

Sudoku Seventeen, by John Austin

115 (4*)

5	4	1	9	8	7	3	6	2
6	2	8	1	3	5	9	4	7
9	3	7	2	4	6	8	5	1
7	1	6	8	2	4	5	9	3
3	8	9	6	5	1	2	7	4
2	5	4	7	9	3	6	1	8
4	6	5	3	7	2	1	8	9
8	7	2	5	1	9	4	3	6
1	9	3	4	6	8	7	2	5

118 (4*)

5	6	9	4	8	7	2	3	1
2	7	8	3	9	1	4	5	6
4	1	3	6	5	2	9	8	7
8	4	2	5	7	9	1	6	3
6	3	7	1	4	8	5	9	2
1	9	5	2	6	3	8	7	4
7	5	1	8	2	6	3	4	9
9	2	4	7	3	5	6	1	8
3	8	6	9	1	4	7	2	5

116 (4*)

3	5	7	9	4	2	1	8	6
4	9	2	6	8	1	3	7	5
1	6	8	5	3	7	2	4	9
6	4	3	7	9	5	8	1	2
2	7	1	8	6	3	9	5	4
9	8	5	2	1	4	7	6	3
7	2	6	3	5	8	4	9	1
5	3	4	1	7	9	6	2	8
8	1	9	4	2	6	5	3	7

119 (4*)

2	7	4	3	8	9	1	5	6
6	1	5	4	7	2	9	3	8
3	8	9	1	6	5	2	7	4
5	3	8	2	1	7	6	4	9
9	4	6	5	3	8	7	2	1
7	2	1	9	4	6	3	8	5
8	9	7	6	2	4	5	1	3
4	5	3	7	9	1	8	6	2
1	6	2	8	5	3	4	9	7

117 (4*)

1	5	8	6	7	3	4	9	2
9	3	2	4	8	5	7	6	1
7	4	6	9	1	2	3	8	5
3	7	4	1	5	6	9	2	8
2	6	1	8	3	9	5	4	7
8	9	5	7	2	4	6	1	3
6	8	3	2	9	7	1	5	4
5	2	9	3	4	1	8	7	6
4	1	7	5	6	8	2	3	9

120 (4*)

4	8	9	1	3	7	5	2	6
1	2	7	5	9	6	4	3	8
6	5	3	8	2	4	7	1	9
9	7	5	2	6	3	1	8	4
3	4	8	9	1	5	6	7	2
2	1	6	4	7	8	3	9	5
8	3	1	6	4	9	2	5	7
7	9	4	3	5	2	8	6	1
5	6	2	7	8	1	9	4	3

Sudoku Seventeen, by John Austin

121 (4*)

1	8	7	6	9	4	3	5	2
3	5	6	8	7	2	9	4	1
2	4	9	3	1	5	8	6	7
6	1	4	9	2	8	7	3	5
8	3	2	4	5	7	6	1	9
7	9	5	1	3	6	4	2	8
9	7	3	2	4	1	5	8	6
4	2	8	5	6	9	1	7	3
5	6	1	7	8	3	2	9	4

124 (4*)

2	9	8	5	3	7	6	1	4
6	5	3	4	1	8	2	7	9
4	1	7	2	9	6	3	5	8
9	4	1	8	6	5	7	3	2
5	7	6	3	4	2	8	9	1
8	3	2	1	7	9	5	4	6
7	2	9	6	5	1	4	8	3
1	8	4	7	2	3	9	6	5
3	6	5	9	8	4	1	2	7

122 (4*)

1	7	3	2	8	4	6	9	5
2	6	8	3	5	9	4	7	1
9	5	4	6	1	7	3	8	2
5	8	6	1	4	3	9	2	7
4	3	1	7	9	2	5	6	8
7	2	9	8	6	5	1	3	4
3	4	2	9	7	1	8	5	6
6	1	7	5	3	8	2	4	9
8	9	5	4	2	6	7	1	3

125 (4*)

2	9	1	5	3	6	4	7	8
4	3	7	2	8	9	6	5	1
5	6	8	7	1	4	9	3	2
8	7	2	9	5	1	3	6	4
9	4	3	8	6	7	1	2	5
6	1	5	3	4	2	8	9	7
3	2	4	6	7	8	5	1	9
7	8	6	1	9	5	2	4	3
1	5	9	4	2	3	7	8	6

123 (4*)

1	4	5	8	9	6	7	3	2
8	6	3	7	2	4	5	1	9
7	2	9	1	3	5	8	4	6
5	8	2	6	1	3	9	7	4
3	1	6	9	4	7	2	5	8
9	7	4	2	5	8	1	6	3
4	9	7	5	6	2	3	8	1
2	3	8	4	7	1	6	9	5
6	5	1	3	8	9	4	2	7

126 (4*)

2	3	9	4	5	8	6	1	7
6	1	5	9	7	3	2	8	4
7	4	8	1	2	6	9	3	5
4	6	3	5	8	7	1	2	9
5	7	2	6	1	9	3	4	8
9	8	1	2	3	4	5	7	6
3	9	4	7	6	1	8	5	2
8	2	6	3	4	5	7	9	1
1	5	7	8	9	2	4	6	3

127 (4*)

2	1	8	9	3	7	4	5	6
4	9	7	5	8	6	3	2	1
6	5	3	2	1	4	8	9	7
3	2	4	6	5	9	7	1	8
5	7	1	3	2	8	6	4	9
9	8	6	7	4	1	5	3	2
8	3	9	1	6	5	2	7	4
1	4	5	8	7	2	9	6	3
7	6	2	4	9	3	1	8	5

130 (4*)

4	8	9	3	6	7	5	2	1
6	1	7	2	9	5	4	3	8
2	5	3	8	1	4	9	6	7
1	3	5	6	2	9	7	8	4
8	9	6	4	7	1	2	5	3
7	4	2	5	8	3	1	9	6
5	6	1	9	4	8	3	7	2
9	7	8	1	3	2	6	4	5
3	2	4	7	5	6	8	1	9

128 (4*)

3	2	9	1	4	6	5	7	8
6	1	7	8	2	5	4	3	9
8	5	4	9	7	3	1	6	2
4	9	8	2	3	1	6	5	7
5	6	1	4	9	7	2	8	3
7	3	2	5	6	8	9	1	4
1	4	3	7	5	2	8	9	6
2	7	5	6	8	9	3	4	1
9	8	6	3	1	4	7	2	5

131 (4*)

4	8	2	6	9	5	3	1	7
5	7	3	2	1	4	9	6	8
1	9	6	3	7	8	5	4	2
6	4	7	1	3	9	8	2	5
3	5	9	8	6	2	1	7	4
2	1	8	5	4	7	6	9	3
9	6	5	7	2	3	4	8	1
8	2	4	9	5	1	7	3	6
7	3	1	4	8	6	2	5	9

129 (4*)

3	1	8	9	5	7	6	2	4
9	6	4	2	1	8	3	5	7
7	5	2	3	6	4	1	9	8
8	2	6	7	9	3	4	1	5
5	7	3	1	4	2	8	6	9
1	4	9	6	8	5	7	3	2
4	9	7	5	3	1	2	8	6
2	3	5	8	7	6	9	4	1
6	8	1	4	2	9	5	7	3

132 (4*)

4	5	7	1	6	3	8	2	9
8	1	9	2	4	7	5	6	3
6	2	3	8	9	5	4	7	1
3	6	2	5	1	4	9	8	7
1	4	8	7	3	9	6	5	2
7	9	5	6	8	2	3	1	4
9	8	6	4	2	1	7	3	5
5	3	1	9	7	6	2	4	8
2	7	4	3	5	8	1	9	6

Sudoku Seventeen, by John Austin

133 (4*)

5	9	8	3	6	7	2	4	1
7	4	3	2	8	1	5	9	6
2	6	1	9	5	4	8	7	3
4	1	7	6	3	5	9	8	2
3	2	6	8	4	9	1	5	7
9	8	5	7	1	2	6	3	4
8	3	4	1	9	6	7	2	5
1	7	9	5	2	3	4	6	8
6	5	2	4	7	8	3	1	9

136 (4*)

6	8	4	2	3	1	5	9	7
7	2	3	5	9	6	1	8	4
1	9	5	4	7	8	2	3	6
3	6	9	1	5	7	4	2	8
5	4	8	9	6	2	3	7	1
2	1	7	8	4	3	6	5	9
4	7	2	6	8	5	9	1	3
9	3	1	7	2	4	8	6	5
8	5	6	3	1	9	7	4	2

134 (4*)

5	2	3	1	8	9	6	4	7
7	6	9	4	3	5	2	8	1
4	1	8	7	2	6	9	5	3
1	7	4	2	6	8	3	9	5
6	3	5	9	4	7	1	2	8
9	8	2	3	5	1	7	6	4
8	9	7	6	1	4	5	3	2
2	4	1	5	9	3	8	7	6
3	5	6	8	7	2	4	1	9

137 (4*)

6	8	5	7	3	1	2	4	9
4	7	9	5	6	2	8	3	1
1	2	3	9	4	8	6	7	5
7	1	6	2	8	5	4	9	3
5	3	8	4	7	9	1	6	2
2	9	4	3	1	6	5	8	7
3	4	1	6	2	7	9	5	8
8	5	7	1	9	4	3	2	6
9	6	2	8	5	3	7	1	4

135 (4*)

6	8	5	4	9	3	7	2	1
4	1	7	8	5	2	9	3	6
2	3	9	7	1	6	4	8	5
7	2	1	3	4	5	6	9	8
3	6	4	2	8	9	1	5	7
5	9	8	6	7	1	2	4	3
9	5	2	1	3	7	8	6	4
8	7	3	9	6	4	5	1	2
1	4	6	5	2	8	3	7	9

138 (4*)

6	3	1	2	5	8	9	4	7
7	9	4	1	6	3	5	2	8
8	2	5	9	4	7	1	6	3
3	4	8	5	7	2	6	9	1
5	1	9	6	8	4	3	7	2
2	6	7	3	1	9	4	8	5
4	5	6	7	2	1	8	3	9
9	8	2	4	3	5	7	1	6
1	7	3	8	9	6	2	5	4

Sudoku Seventeen, by John Austin

139 (4*)

7	9	8	6	5	1	4	3	2
2	5	6	8	4	3	1	7	9
1	4	3	9	7	2	8	6	5
3	6	9	7	1	8	5	2	4
4	2	1	5	3	6	9	8	7
8	7	5	4	2	9	3	1	6
9	1	7	2	8	5	6	4	3
5	8	4	3	6	7	2	9	1
6	3	2	1	9	4	7	5	8

140 (4*)

9	6	1	2	5	3	8	4	7
4	8	5	9	1	7	3	6	2
3	2	7	8	6	4	1	9	5
2	3	8	6	7	1	9	5	4
7	5	4	3	9	8	2	1	6
1	9	6	5	4	2	7	3	8
6	7	9	1	8	5	4	2	3
5	4	3	7	2	9	6	8	1
8	1	2	4	3	6	5	7	9

141 (4*)

4	5	2	9	6	3	7	8	1
8	1	7	2	4	5	9	3	6
9	3	6	7	8	1	5	2	4
7	6	8	1	5	9	3	4	2
3	4	1	6	7	2	8	5	9
2	9	5	8	3	4	1	6	7
6	2	3	5	1	7	4	9	8
5	7	9	4	2	8	6	1	3
1	8	4	3	9	6	2	7	5

142 (4*)

7	1	3	9	5	8	2	6	4
9	2	6	4	3	1	8	7	5
5	8	4	6	2	7	1	9	3
3	9	1	7	8	6	4	5	2
6	4	7	5	1	2	3	8	9
8	5	2	3	4	9	6	1	7
1	3	9	8	7	4	5	2	6
4	7	8	2	6	5	9	3	1
2	6	5	1	9	3	7	4	8

143 (4*)

6	4	7	3	1	8	9	2	5
1	2	3	4	9	5	8	7	6
5	8	9	7	2	6	4	3	1
4	7	6	9	5	2	3	1	8
9	5	8	1	3	7	2	6	4
3	1	2	6	8	4	7	5	9
8	3	5	2	4	1	6	9	7
7	9	1	8	6	3	5	4	2
2	6	4	5	7	9	1	8	3

144 (4*)

4	6	5	9	2	1	3	8	7
8	7	3	5	4	6	9	1	2
2	9	1	3	8	7	6	4	5
7	1	4	2	6	9	8	5	3
5	3	8	7	1	4	2	6	9
6	2	9	8	3	5	1	7	4
1	5	2	6	7	3	4	9	8
3	4	7	1	9	8	5	2	6
9	8	6	4	5	2	7	3	1

Sudoku Seventeen, by John Austin

145 (4*)

4	3	9	1	2	5	8	7	6
2	8	6	4	7	3	9	1	5
5	7	1	8	9	6	3	4	2
8	4	3	9	6	2	7	5	1
9	6	5	7	3	1	2	8	4
7	1	2	5	4	8	6	3	9
1	9	8	6	5	7	4	2	3
3	5	4	2	8	9	1	6	7
6	2	7	3	1	4	5	9	8

148 (4*)

4	7	6	3	1	9	5	8	2
2	8	3	4	6	5	7	1	9
5	1	9	2	7	8	3	4	6
9	6	8	7	2	3	4	5	1
1	5	7	9	8	4	2	6	3
3	4	2	1	5	6	9	7	8
8	2	1	5	3	7	6	9	4
6	9	5	8	4	2	1	3	7
7	3	4	6	9	1	8	2	5

146 (4*)

9	5	8	7	2	3	6	4	1
2	3	4	5	6	1	8	7	9
6	7	1	4	9	8	3	2	5
7	8	9	6	1	4	2	5	3
5	1	2	3	8	7	4	9	6
3	4	6	2	5	9	7	1	8
1	6	5	8	4	2	9	3	7
8	2	3	9	7	5	1	6	4
4	9	7	1	3	6	5	8	2

149 (4*)

2	5	6	4	7	9	1	8	3
3	9	7	5	1	8	4	2	6
8	4	1	2	3	6	5	7	9
7	2	8	9	4	3	6	1	5
5	3	9	1	6	7	2	4	8
1	6	4	8	2	5	3	9	7
4	8	3	7	5	2	9	6	1
9	1	5	6	8	4	7	3	2
6	7	2	3	9	1	8	5	4

147 (4*)

3	2	6	8	7	5	1	9	4
5	9	8	4	6	1	7	2	3
4	7	1	3	9	2	5	8	6
7	1	4	2	8	6	9	3	5
9	5	3	1	4	7	2	6	8
6	8	2	5	3	9	4	1	7
2	4	5	6	1	3	8	7	9
1	3	7	9	5	8	6	4	2
8	6	9	7	2	4	3	5	1

150 (4*)

8	2	4	1	5	9	6	7	3
5	6	1	2	7	3	8	9	4
3	9	7	4	6	8	2	5	1
6	3	2	5	9	1	4	8	7
4	7	5	8	3	6	1	2	9
1	8	9	7	4	2	5	3	6
9	1	3	6	8	5	7	4	2
2	4	8	9	1	7	3	6	5
7	5	6	3	2	4	9	1	8

Sudoku Seventeen, by John Austin

151 (5*)

5	8	7	9	3	4	1	2	6
3	9	6	1	2	5	8	7	4
1	2	4	7	8	6	5	9	3
4	3	9	6	1	8	2	5	7
8	7	5	3	4	2	6	1	9
6	1	2	5	7	9	3	4	8
2	5	8	4	6	7	9	3	1
9	4	1	8	5	3	7	6	2
7	6	3	2	9	1	4	8	5

154 (5*)

3	2	9	6	8	7	4	5	1
6	7	1	2	4	5	8	9	3
4	5	8	3	1	9	2	6	7
1	8	6	5	9	3	7	4	2
2	3	5	4	7	1	9	8	6
9	4	7	8	2	6	3	1	5
8	1	3	7	6	4	5	2	9
5	6	2	9	3	8	1	7	4
7	9	4	1	5	2	6	3	8

152 (5*)

1	3	9	8	6	9	2	5	4
8	5	2	9	4	1	3	7	6
4	6	7	3	2	5	1	8	9
6	8	5	2	9	4	7	1	3
3	9	1	6	5	7	4	2	8
2	7	4	1	8	3	6	9	5
9	2	3	5	1	6	8	4	7
5	4	8	7	3	2	9	6	1
7	1	6	4	9	8	5	3	2

155 (5*)

5	3	9	8	2	1	6	7	4
6	2	1	4	3	7	5	8	9
4	8	7	9	6	5	2	3	1
3	9	4	2	7	8	1	6	5
1	7	8	5	4	6	9	2	3
2	6	5	3	1	9	8	4	7
9	4	6	7	5	2	3	1	8
7	5	2	1	8	3	4	9	6
8	1	3	6	9	4	7	5	2

153 (5*)

7	8	2	5	6	1	3	4	9
9	6	1	2	4	3	5	8	7
4	5	3	8	7	9	2	1	6
3	2	6	9	8	4	1	7	5
8	1	9	3	5	7	4	6	2
5	4	7	1	2	6	8	9	3
6	3	8	7	1	5	9	2	4
1	7	5	4	9	2	6	3	8
2	9	4	6	3	8	7	5	1

156 (5*)

4	7	3	6	2	1	5	8	9
8	2	9	5	3	4	7	6	1
1	5	6	7	8	9	3	2	4
6	1	2	4	9	3	8	7	5
9	8	5	2	1	7	4	3	6
3	4	7	8	6	5	9	1	2
2	3	8	9	5	6	1	4	7
7	9	1	3	4	2	6	5	8
5	6	4	1	8	7	2	9	3

Sudoku Seventeen, by John Austin

157 (5*)

4	5	2	8	9	7	3	6	1
7	6	8	3	1	4	2	5	9
3	1	9	2	5	6	4	8	7
9	3	5	4	8	1	6	7	2
6	4	1	7	2	9	5	3	8
8	2	7	6	3	5	1	9	4
1	7	3	9	6	2	8	4	5
5	8	4	1	7	3	9	2	6
2	9	6	5	4	8	7	1	3

160 (5*)

8	9	7	6	5	2	3	1	4
3	4	2	1	7	8	6	5	9
6	5	1	9	3	4	2	8	7
9	3	6	4	8	1	5	7	2
2	8	4	7	6	5	1	9	3
1	7	5	2	9	3	8	4	6
5	2	9	3	1	7	4	6	8
7	1	3	8	4	6	9	2	5
4	6	8	5	2	9	7	3	1

158 (5*)

1	2	8	3	4	5	6	9	7
7	5	9	8	1	6	2	3	4
6	4	3	2	9	7	8	5	1
2	9	6	5	7	4	3	1	8
4	3	1	6	8	9	7	2	5
5	8	7	1	2	3	9	4	6
8	6	2	4	3	1	5	7	9
9	1	5	7	6	2	4	8	3
3	7	4	9	5	8	1	6	2

161 (5*)

4	6	8	3	7	2	9	1	5
3	7	1	8	5	9	4	2	6
5	2	9	6	1	4	7	8	3
8	1	4	5	9	7	6	3	2
2	5	7	1	3	6	8	4	9
6	9	3	2	4	8	5	7	1
9	8	2	4	6	3	1	5	7
7	3	5	9	8	1	2	6	4
1	4	6	7	2	5	3	9	8

159 (5*)

5	9	2	7	3	4	8	6	1
3	7	6	1	9	8	2	5	4
1	8	4	6	2	5	7	9	3
9	1	8	3	5	2	6	4	7
4	5	7	9	8	6	1	3	2
2	6	3	4	7	1	9	8	5
6	2	1	5	4	9	3	7	8
7	4	9	8	1	3	5	2	6
8	3	5	2	6	7	4	1	9

162 (5*)

3	4	6	7	1	8	5	2	9
8	7	1	5	2	9	4	6	3
9	5	2	3	4	6	7	8	1
6	1	9	2	7	4	3	5	8
4	3	7	9	8	5	2	1	6
2	8	5	1	6	3	9	4	7
1	2	4	8	3	7	6	9	5
5	6	3	4	9	1	8	7	2
7	9	8	6	5	2	1	3	4

Sudoku Seventeen, by John Austin

163 (5*)

7	6	3	4	1	2	5	9	8
9	4	1	8	5	6	7	3	2
2	5	8	3	7	9	4	6	1
6	7	4	2	3	1	9	8	5
8	2	9	5	6	4	3	1	7
1	3	5	9	8	7	6	2	4
3	8	7	6	2	5	1	4	9
5	9	6	1	4	8	2	7	3
4	1	2	7	9	3	8	5	6

166 (5*)

4	8	9	6	2	5	7	1	3
7	6	1	4	8	3	5	2	9
5	3	2	9	1	7	4	6	8
3	2	6	5	7	9	1	8	4
9	4	5	8	6	1	3	7	2
1	7	8	2	3	4	6	9	5
6	5	7	3	9	8	2	4	1
2	9	3	1	4	6	8	5	7
8	1	4	7	5	2	9	3	6

164 (5*)

8	1	6	3	2	5	9	4	7
7	9	4	8	1	6	3	5	2
2	5	3	9	7	4	8	1	6
5	2	1	7	6	3	4	9	8
6	3	7	4	9	8	1	2	5
4	8	9	1	5	2	6	7	3
1	4	2	6	8	7	5	3	9
3	7	8	5	4	9	2	6	1
9	6	5	2	3	1	7	8	4

167 (5*)

7	1	4	9	3	2	5	8	6
6	9	2	5	1	8	3	7	4
8	5	3	7	4	6	1	2	9
1	2	7	4	6	5	8	9	3
3	4	5	8	9	1	7	6	2
9	6	8	2	7	3	4	5	1
5	3	6	1	8	9	2	4	7
4	8	1	6	2	7	9	3	5
2	7	9	3	5	4	6	1	8

165 (5*)

7	3	8	9	2	5	4	6	1
9	6	5	4	7	1	2	8	3
1	4	2	3	6	8	5	7	9
2	9	3	8	5	4	6	1	7
6	8	7	1	9	2	3	5	4
4	5	1	7	3	6	9	2	8
8	2	9	5	4	7	1	3	6
3	7	6	2	1	9	8	4	5
5	1	4	6	8	3	7	9	2

168 (5*)

9	2	5	8	4	6	1	7	3
6	1	8	3	9	7	5	2	4
4	3	7	1	5	2	6	9	8
5	9	3	2	7	1	8	4	6
8	4	2	9	6	5	7	3	1
7	6	1	4	3	8	2	5	9
2	7	4	6	8	3	9	1	5
3	5	6	7	1	9	4	8	2
1	8	9	5	2	4	3	6	7

Sudoku Seventeen, by John Austin

169 (5*)

5	3	7	9	6	8	1	2	4
8	2	4	7	3	1	6	9	5
9	1	6	4	5	2	8	3	7
1	7	5	2	9	6	4	8	3
3	4	9	1	8	5	7	6	2
2	6	8	3	4	7	5	1	9
7	9	2	8	1	4	3	5	6
4	5	1	6	2	3	9	7	8
6	8	3	5	7	9	2	4	1

172 (5*)

9	1	3	6	7	8	4	2	5
5	8	6	4	2	3	9	1	7
2	7	4	9	1	5	8	6	3
8	4	1	5	6	7	3	9	2
6	2	7	3	9	4	1	5	8
3	9	5	1	8	2	7	4	6
7	6	9	2	3	1	5	8	4
4	3	2	8	5	9	6	7	1
1	5	8	7	4	6	2	3	9

170 (5*)

2	6	3	7	5	8	9	1	4
4	8	1	6	9	3	7	2	5
9	5	7	4	1	2	3	8	6
3	9	5	8	4	1	6	7	2
6	1	2	3	7	9	5	4	8
8	7	4	5	2	6	1	9	3
5	4	9	2	3	7	8	6	1
1	2	8	9	6	5	4	3	7
7	3	6	1	8	4	2	5	9

173 (5*)

6	2	7	5	4	1	3	8	9
4	8	1	7	9	3	5	2	6
5	9	3	2	8	6	7	1	4
1	7	4	9	5	8	6	3	2
8	5	9	3	6	2	1	4	7
3	6	2	4	1	7	8	9	5
2	3	5	8	7	4	9	6	1
7	1	8	6	2	9	4	5	3
9	4	6	1	3	5	2	7	8

171 (5*)

7	1	2	8	4	9	6	5	3
9	5	4	6	2	3	7	1	8
6	3	8	1	7	5	4	2	9
1	8	6	4	9	7	5	3	2
3	7	5	2	1	6	8	9	4
4	2	9	3	5	8	1	6	7
8	9	7	5	6	2	3	4	1
2	6	1	7	3	4	9	8	5
5	4	3	9	8	1	2	7	6

174 (5*)

3	1	9	2	7	6	8	5	4
7	2	4	8	5	3	9	6	1
6	5	8	1	9	4	2	7	3
8	4	7	6	1	5	3	9	2
5	6	2	3	8	9	1	4	7
1	9	3	4	2	7	6	8	5
2	8	6	7	4	1	5	3	9
9	7	1	5	3	8	4	2	6
4	3	5	9	6	2	7	1	8

Sudoku Seventeen, by John Austin

175 (5*)

1	3	7	4	2	9	5	8	6
9	2	5	8	6	1	4	3	7
4	6	8	7	3	5	1	9	2
6	9	4	3	1	7	2	5	8
8	1	2	5	4	6	3	7	9
7	5	3	2	9	8	6	1	4
2	7	9	1	5	4	8	6	3
5	4	6	9	8	3	7	2	1
3	8	1	6	7	2	9	4	5

178 (5*)

8	6	1	5	2	9	3	4	7
3	7	9	8	6	4	5	1	2
2	4	5	7	3	1	6	8	9
9	3	4	6	1	7	8	2	5
7	1	2	3	8	5	4	9	6
6	5	8	9	4	2	7	3	1
1	8	7	2	5	3	9	6	4
4	9	6	1	7	8	2	5	3
5	2	3	4	9	6	1	7	8

176 (5*)

1	4	5	2	3	7	8	6	9
7	3	8	6	9	5	1	2	4
6	9	2	4	1	8	5	7	3
5	6	9	1	8	4	2	3	7
4	8	7	3	5	2	9	1	6
3	2	1	9	7	6	4	5	8
8	5	6	7	4	1	3	9	2
9	7	4	5	2	3	6	8	1
2	1	3	8	6	9	7	4	5

179 (5*)

7	8	3	9	4	1	5	2	6
1	5	9	2	6	3	7	4	8
4	2	6	8	5	7	1	3	9
3	6	4	5	9	8	2	7	1
2	7	8	3	1	6	4	9	5
5	9	1	4	7	2	6	8	3
6	4	5	7	3	9	8	1	2
8	3	7	1	2	5	9	6	4
9	1	2	6	8	4	3	5	7

177 (5*)

6	5	7	9	4	2	3	1	8
8	1	4	7	3	6	9	5	2
3	9	2	1	5	8	4	6	7
4	6	9	5	2	1	7	8	3
7	2	3	8	6	9	5	4	1
1	8	5	3	7	4	6	2	9
5	3	6	2	8	7	1	9	4
2	4	1	6	9	3	8	7	5
9	7	8	4	1	5	2	3	6

180 (5*)

9	8	4	5	1	7	6	2	3
5	6	3	8	2	9	7	4	1
1	7	2	3	4	6	8	5	9
8	5	7	1	6	2	9	3	4
6	4	1	7	9	3	2	8	5
3	2	9	4	5	8	1	7	6
7	1	8	9	3	4	5	6	2
4	9	6	2	7	5	3	1	8
2	3	5	6	8	1	4	9	7

Sudoku Seventeen, by John Austin

181 (5*)

4	9	1	5	7	3	2	8	6
2	6	3	4	8	9	1	5	7
8	7	5	2	6	1	3	9	4
9	2	4	6	1	7	5	3	8
3	8	7	9	5	2	6	4	1
1	5	6	8	3	4	9	7	2
6	3	8	1	4	5	7	2	9
5	4	2	7	9	6	8	1	3
7	1	9	3	2	8	4	6	5

184 (5*)

6	5	7	1	9	4	8	3	2
1	3	8	5	2	6	7	4	9
2	9	4	8	3	7	5	1	6
5	4	2	6	7	3	9	8	1
3	8	1	2	5	9	6	7	4
7	6	9	4	1	8	2	5	3
8	2	6	7	4	1	3	9	5
9	1	5	3	8	2	4	6	7
4	7	3	9	6	5	1	2	8

182 (5*)

6	9	8	5	2	1	7	4	3
5	7	3	9	4	8	6	1	2
1	2	4	6	7	3	5	9	8
8	5	2	1	9	7	3	6	4
3	1	7	4	6	5	2	8	9
4	6	9	8	3	2	1	7	5
7	3	1	2	8	9	4	5	6
2	8	6	7	5	4	9	3	1
9	4	5	3	1	6	8	2	7

185 (5*)

1	8	4	9	6	7	3	5	2
2	3	5	1	8	4	7	6	9
7	6	9	2	5	3	4	1	8
9	7	3	8	1	2	6	4	5
4	2	6	7	9	5	1	8	3
5	1	8	3	4	6	2	9	7
6	9	2	5	3	1	8	7	4
8	4	7	6	2	9	5	3	1
3	5	1	4	7	8	9	2	6

183 (5*)

7	6	9	3	5	4	2	8	1
8	4	5	1	2	6	9	7	3
2	1	3	7	8	9	5	6	4
1	7	8	6	3	2	4	5	9
5	3	4	9	7	1	6	2	8
9	2	6	8	4	5	3	1	7
4	9	1	5	6	7	8	3	2
3	5	2	4	1	8	7	9	6
6	8	7	2	9	3	1	4	5

186 (5*)

9	2	7	6	1	4	5	3	8
6	3	8	5	9	2	4	7	1
5	4	1	8	7	3	9	6	2
1	6	2	9	4	8	7	5	3
7	8	5	1	3	6	2	4	9
4	9	3	7	2	5	1	8	6
3	5	4	2	6	1	8	9	7
8	1	9	3	5	7	6	2	4
2	7	6	4	8	9	3	1	5

Sudoku Seventeen, by John Austin

187 (5*)

1	6	9	5	3	8	7	4	2
7	8	4	6	2	9	5	1	3
2	5	3	1	4	7	9	8	6
3	1	6	4	7	5	2	9	8
5	2	8	3	9	6	4	7	1
4	9	7	2	8	1	6	3	5
6	3	2	7	1	4	8	5	9
9	4	5	8	6	3	1	2	7
8	7	1	9	5	2	3	6	4

190 (5*)

2	8	4	6	9	3	5	1	7
3	6	1	5	2	7	9	8	4
7	9	5	8	4	1	2	6	3
6	4	8	1	7	2	3	9	5
9	5	7	3	6	4	8	2	1
1	2	3	9	8	5	4	7	6
4	3	6	2	1	8	7	5	9
5	1	2	7	3	9	6	4	8
8	7	9	4	5	6	1	3	2

188 (5*)

1	4	6	8	3	9	7	5	2
8	7	9	2	4	5	1	3	6
2	3	5	6	1	7	8	9	4
7	6	3	5	2	1	9	4	8
4	8	2	7	9	3	5	6	1
9	5	1	4	6	8	3	2	7
6	1	8	9	5	2	4	7	3
3	9	4	1	7	6	2	8	5
5	2	7	3	8	4	6	1	9

191 (5*)

7	1	5	6	8	2	3	4	9
6	2	8	3	9	4	1	7	5
9	4	3	1	5	7	2	8	6
5	7	2	9	1	6	4	3	8
8	3	1	4	2	5	9	6	7
4	9	6	8	7	3	5	2	1
3	5	7	2	6	9	8	1	4
1	6	4	5	3	8	7	9	2
2	8	9	7	4	1	6	5	3

189 (5*)

6	1	9	5	2	7	4	8	3
7	4	8	1	3	9	2	6	5
3	5	2	8	6	4	9	1	7
1	2	5	9	8	3	6	7	4
8	7	6	2	4	5	1	3	9
4	9	3	6	7	1	8	5	2
5	3	1	4	9	6	7	2	8
2	6	4	7	5	8	3	9	1
9	8	7	3	1	2	5	4	6

192 (5*)

8	5	7	3	9	1	6	2	4
6	3	9	2	4	5	1	7	8
1	2	4	7	8	6	9	3	5
7	1	5	4	6	2	8	9	3
3	4	8	1	7	9	5	6	2
2	9	6	8	5	3	7	4	1
9	7	2	5	3	8	4	1	6
5	6	1	9	2	4	3	8	7
4	8	3	6	1	7	2	5	9

Sudoku Seventeen, by John Austin

193 (6*)

5	8	6	7	3	9	2	4	1
3	4	2	8	1	6	9	7	5
9	1	7	5	2	4	8	3	6
7	3	1	2	4	5	6	8	9
2	5	8	6	9	7	4	1	3
6	9	4	1	8	3	7	5	2
8	6	3	4	5	2	1	9	7
4	2	5	9	7	1	3	6	8
1	7	9	3	6	8	5	2	4

196 (6*)

5	7	2	1	6	4	3	8	9
1	6	4	9	8	3	5	2	7
8	9	3	5	2	7	1	6	4
7	2	9	8	5	1	4	3	6
6	3	8	4	9	2	7	5	1
4	5	1	3	7	6	8	9	2
3	4	5	2	1	9	6	7	8
9	8	7	6	4	5	2	1	3
2	1	6	7	3	8	9	4	5

194 (6*)

9	8	7	6	3	1	4	5	2
1	6	4	7	2	5	8	3	9
2	3	5	4	8	9	6	1	7
5	4	1	8	9	7	3	2	6
8	2	6	1	4	3	9	7	5
3	7	9	2	5	6	1	4	8
6	9	2	3	7	4	5	8	1
7	1	3	5	6	8	2	9	4
4	5	8	9	1	2	7	6	3

197 (6*)

6	8	5	2	4	1	9	7	3
3	4	7	9	5	6	1	2	8
1	2	9	3	8	7	6	5	4
5	1	6	8	3	2	7	4	9
8	3	2	4	7	9	5	1	6
7	9	4	6	1	5	8	3	2
2	7	8	1	6	4	3	9	5
4	6	1	5	9	3	2	8	7
9	5	3	7	2	8	4	6	1

195 (6*)

4	2	6	9	3	1	5	8	7
9	5	3	8	7	6	2	4	1
1	7	8	2	5	4	6	9	3
6	9	1	3	2	5	8	7	4
3	8	5	7	4	9	1	6	2
2	4	7	6	1	8	9	3	5
7	6	4	1	8	2	3	5	9
5	1	9	4	6	3	7	2	8
8	3	2	5	9	7	4	1	6

198 (6*)

3	8	9	7	6	4	2	5	1
4	7	2	8	1	5	9	3	6
1	5	6	9	3	2	4	7	8
8	9	7	6	2	1	3	4	5
6	3	1	4	5	9	8	2	7
5	2	4	3	7	8	1	6	9
9	1	5	2	4	6	7	8	3
7	4	8	5	9	3	6	1	2
2	6	3	1	8	7	5	9	4

Sudoku Seventeen, by John Austin

199 (6*)

3	5	1	8	7	9	4	2	6
4	9	2	6	1	3	5	7	8
7	8	6	4	2	5	3	9	1
2	1	8	7	9	4	6	3	5
9	6	3	5	8	1	7	4	2
5	4	7	3	6	2	8	1	9
1	7	5	9	3	6	2	8	4
6	3	9	2	4	8	1	5	7
8	2	4	1	5	7	9	6	3

202 (6*)

6	9	1	3	7	4	5	2	8
7	3	4	8	5	2	1	9	6
2	5	8	6	9	1	4	7	3
3	1	2	5	8	7	9	6	4
5	6	7	4	3	9	8	1	2
8	4	9	1	2	6	3	5	7
9	8	6	2	1	3	7	4	5
4	7	5	9	6	8	2	3	1
1	2	3	7	4	5	6	8	9

200 (6*)

6	8	9	2	5	3	4	1	7
5	2	4	6	7	1	3	9	8
7	1	3	8	9	4	2	6	5
4	7	6	1	3	9	8	5	2
9	3	8	5	2	7	1	4	6
2	5	1	4	6	8	7	3	9
1	6	7	9	4	2	5	8	3
3	4	5	7	8	6	9	2	1
8	9	2	3	1	5	6	7	4

203 (6*)

2	9	4	3	1	6	5	7	8
7	1	5	4	2	8	9	6	3
3	8	6	7	9	5	2	4	1
6	4	2	8	3	9	1	5	7
9	7	1	5	6	2	8	3	4
5	3	8	1	4	7	6	2	9
1	2	7	9	5	4	3	8	6
4	5	9	6	8	3	7	1	2
8	6	3	2	7	1	4	9	5

201 (6*)

2	1	5	8	4	6	7	3	9
9	6	4	7	5	3	8	2	1
8	3	7	9	2	1	5	6	4
1	8	9	2	6	5	4	7	3
6	5	3	4	1	7	9	8	2
7	4	2	3	8	9	1	5	6
4	7	6	1	3	8	2	9	5
3	2	8	5	9	4	6	1	7
5	9	1	6	7	2	3	4	8

204 (6*)

1	6	2	7	9	8	3	4	5
9	8	4	3	5	2	1	6	7
3	7	5	4	1	6	2	9	8
2	5	6	8	7	9	4	3	1
4	9	7	1	3	5	8	2	6
8	1	3	6	2	4	7	5	9
6	3	8	9	4	1	5	7	2
5	4	1	2	6	7	9	8	3
7	2	9	5	8	3	6	1	4

www.ingramcontent.com/pod-product-compliance
Lightning Source LLC
Chambersburg PA
CBHW051907170526
45168CB00001B/277